# MODDESHALL HYDROPOWER

## THE PAST AND POTENTIAL OF HYDROPOWER IN THE MODDESHALL VALLEY

James R Warren

First Published in the United Kingdom in 2022 by Midland Tutorial Productions

First Edition:   15 July 2022

File Prefix Code:       MODRMILL

ISBN 978 1 7396296 2 5

Midland Tutorial Productions Publishers
31 Victoria Avenue
Bloxwich
Walsall
WS3 3HS
United Kingdom

# M I DLAND
# T UTORIAL

# MODDESHALL HYDROPOWER

## THE PAST AND POTENTIAL OF HYDROPOWER IN THE MODDESHALL VALLEY

First Edition
2022

**James R Warren**

MIDLAND TUTORIAL PRODUCTIONS
BLOXWICH

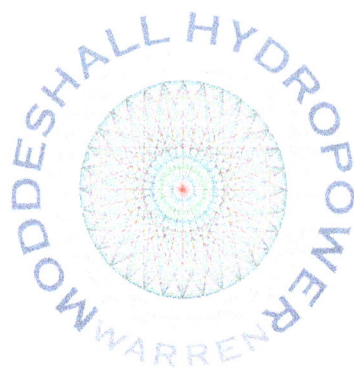

MODDESHALL HYDROPOWER
WARREN

# To The Glory of The Loving God

# Who Made Our Minds Free

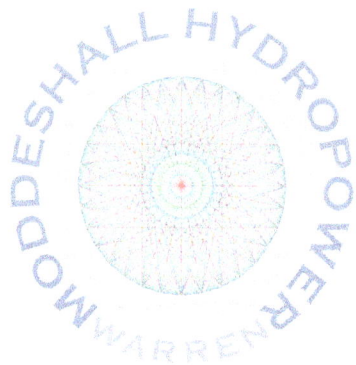

# TABLE OF CONTENTS

# PART IV

# PART V

# PART VI

# TECHNICAL APPENDIX A

MODRMILL
Moddeshall Hydropower     Page 9 of 126
James R Warren     19:03 Sunday, 19 June 2022

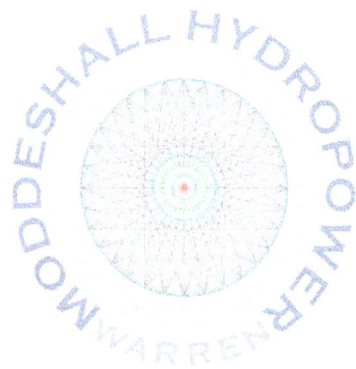

# The Past and Potential of
# Hydropower in the Moddeshall Valley

by
*James R Warren BSc MSc PhD PGCE*

## PART I
## A WALK THROUGH THE MODDESHALL RAVINES

They say that a gigawatt is lurking unloved and long lorn amidst the British countryside, along the gentle falls and placid reaches of our streams. Small beer maybe for a nation that consumes seventy-six, but, hey!, Sir Richard Branson is worth £1.6billion and right now I would find a seventy-sixth part of that sum useful.

The Moddeshall Valley, or more properly the Valley of the Scotch Brook, is a wooded defile in North Staffordshire, a place in Central England. It is only some five miles long, and is incised about 586-296=290 feet ( 88.4 meters ) into soft Permo-Triassic sandstone and, locally, Carboniferous rocks.

The Scotch Brook has two Eastern spur-streams, the Moddeshall Brook and the Cotwalton Brook. The Scotch Brook tends to flow Southwest and contributes to the River Trent at Stone.

Since the Early Middle Ages these streams, especially the two former, have been intensively developed with watermills. Along the Scotch Brook and the Moddeshall Brook south of their confluence, the watermills have cascaded, taking their head from the tail of the mill above. The total fall from Boar Mill Pond to Stone Mill is 530-296=234 feet ( 71.3 meters ) in a distance of about 2.5 miles ( 4.02 kms ). In this distance twelve mills are known to history, of which six survive, none of which are in use. There was a very complex co-ordinated interplay of Splashy Mill and Ochre Mill at the confluence of the Scotch and Moddeshall Brooks.

Originally, the mills ground corn ( i.e. wheat ) or oats, but with the industrialisation of the eighteenth century all but the lowest converted to the preparation of potters' supplies: Tissue paper, linseed oil, Cornish stone, and, predominantly, flint and bone for the bone china industry of nearby Longton.

The Scotch Brook, and especially its Moddeshall Brook headwater, are noted for their reliable and steady celerity, a feature rare in the English Lowlands. This predictability of water availability is due in large measure to the source of the Moddeshall Brook in sandstone springs near to Idle Rocks in the vicinity of SJ930370.

The Scotch Brook is, however, but a shadow of its former self. During the last hundred years abstraction of water for urban supply has reduced flow to such an extent that The Environment Agency categorises the little river as overabstracted and categorically proscribes any new removal.

The existing mills are all equipped with Roman wheels, either of the overshot or high breast designs. The exception is the lowest, Stone Mill, which at closure was found to be equipped with a Fourneyron turbine of early design, possibly of the early 1860's[1].

The valleys are unsuitable for canal navigation and have never accommodated railways.

The morning of Thursday 10th September 2009 dawned still and cloudless in Western Mercia, though a little cumulus rose by noon. There was no mist, but a light haze through which The Wrekin mountain, an isolated and long extinct volcano, could clearly be seen from Moddeshall at a range of twenty-five miles.

It had not rained for some seven days, though the previous month had been unusually wet.

At 1000BST I parked in Cross Street, a quiet terrace in the delightful nineteenth-century town of Stone. The Longton Road A620 follows the valley of the Scotch Brook and its Western side has a narrow tarmacadamed sidewalk as far as a bridge above Wetmore Mill at SJ916360.

First I walked to Coppice Mill at SJ908356. This is a dilapidated but well-roofed shed having a battered timber pitchback wheel with iron shrouds. The soal appears to be missing. The mill is in the grounds of a private dwelling. Built in 1720 as a paper mill it was by 1799 in the possession of Henry Fourdrinier who in 1807 perfected his process for continuous paper production. He applied this process to the production of tissue paper for the transference of artwork onto china pottery. By 1860 Coppice Mill had converted to the wet grinding of flint for bone china production, and continued as a flint mill until closure in 1953. Moddeshall Valley mills are equipped with open tanks for settling ground slurries. These are locally called arks. One of the Coppice Mill arks has been converted to a swimming pool.

**Coppice Mill**

Four hundred meters further along the pavement I walked a little along Nicholls Lane west of the Scotch Brook to picture Hayes Mill ( SJ912350 ) that is now also a dwelling. It has preserved its short drying-kiln chimney and several monitor-roofed ark houses. Hayes Mill was active from 1750 to 1966 and was a flint mill throughout.

**Hayes Mill**

Gently climbing the Scotch Brook valley floor, the A620 enters a dark wooded ravine overshadowed by elegant beech hangers. After a rocky chicane, where trucks have chattered the naked rock cuttings, I reached the magnificent whitewashed Ivy Mill at SJ916355. This is the largest mill complex in the valley and preserves its fine brick smokestack, rising romantically through the tall trees. It is a vision of Staffordshire industrial Arcady of the kind celebrated in a painting by Bonington and Holland, and now sadly almost extinct. Ivy Mill started in 1740 as an oil mill, but by mid-Victorian times was grinding flint. At its closure in 1966 it had converted to bone grinding, also for the Stoke china-pot trade, and subsequently was a warehouse for ten years or so. It seems to now be a dwelling and depot for a local irrigation firm.

Five hundred meters further north Wetmore Mill ( SJ916360 ) survives as a dwelling, but, as with Ivy Mill, none of the movement is visible from the public highway. Wetmore Mill started in 1763 as an oil and flint mill and continued as such throughout the nineteenth century, but had become a bone mill by the time of its closure in 1960.

Just above Wetmore Mill I lost the tarmacadamed sidewalk and became exposed to the full impetuosity of the Longton Road truck traffic. It was prudent to cross to the Eastern side of the carriageway to face the onslaught, where another hundred meters brought me to the gate of the valley's preserved mill complex at Mosty Lee ( SJ918362 ).

**Ivy Mill**

**Mosty Lee Mill**

**The Wheel at Mosty Lee Mill**

Despite the use of public money in the restoration and maintenance of this monument, visitors are strongly discouraged.

Mosty Lee Mill was established as a corn mill in 1716, antedating the British china industry. By 1860 it was a flint mill, and finished its career in 1961 as a flint and ceramics color mill.

After another dangerous half-kilometer northward walk along the A620, I was able to turn East down a narrow and silent country lane heading for Moddeshall village.

After a short downhill stretch I crossed the Scotch Brook and entered the catchment of the Moddeshall Brook and its gentle grassy valley. The Moddeshall Brook is really a fast-flowing canalised lade, showing signs of reversion to a natural stream over much of its course.

At SJ919366 is the impressive surviving building of Splashy Mill complete with its long-still wheel. It appears to remain as an outbuilding in the pleasant back garden of a modern bungalow. Splashy Mill opened in 1752 as a corn mill, but it too had joined the flint-grinding trade by mid-Victorian times. At the time of closure in 1958 it was a bone mill.

Splashy Mill Pond is not visible from the public road, at least not in the luxuriance of summer, but the lane continues uphill past flanking trees to an open section around SJ921366.

The leat flows briskly at the Northern side of the lane for the next hundred meters and this is the site of Point A at which I undertook a cursory hydrometric survey of the Moddershall Brook, which I shall discuss in more detail later.

**Splashy Mill**

**The Moddershall Brook above Point A**

Proceeding after noon to the hamlet of Moddeshall I was able to inspect the extant and extensive mill pond of the now demolished Boar Mill that dominated the village between 1798 and 1954. This too started as a corn mill, but converted to flint grinding at some early opportunity, maybe at the repeal of The Corn Laws in 1846. The millpond, and the tavern that overlooks it are currently ( September 2009 ) being renovated and remodelled.

**Boar Mill Pond in Moddeshall**

In the full heat of the noontide sun I now had a largely featureless but mercifully downhill walk through the back lanes to Stone.

On the way I had an opportunity to snap a half-decent picture of the soft Permo-Triassic sandstone in a road cutting near to Wood House approximately at SJ919354.

After regaining the A620, I walked off down Nanny Goat Lane to visit the former Stone Mill on the Northern flank of Stone town center. This is the lowest of the surviving Scotch Brook mills and is a splendid seven-bay four-storey Georgian brick edifice that was a corn or corn and oat mill throughout its 201-year working life from 1775 to 1976.

It was at Stone Mill, in 1887, that Richard Smith developed the patent wheatgerm flour that is used to make the "Hovis" brand of wholemeal bread, still a staple of the British diet. The mill is now a restaurant, but at the end of its working life was found to mount an early Fourneyron turbine that has now been restored. Across the brook is a complex of derelict Georgian industrial buildings of unknown application.

Stone was long a center of the bitter beer brewing industry. That depended upon clean water, hops, grain and swift, smooth transport.

At the conclusion of this visit I returned to Cross Street and my car. The walk is some 5.67 miles long, occupies a leisurely four hours and includes roughly three hundred feet of ascent. Try to pick a day when the truck drivers are on strike.

**Permo-Triassic Country Rock East of Ivy Mill**

**The Scotch Brook at Stone Mill**

**Stone Mill**

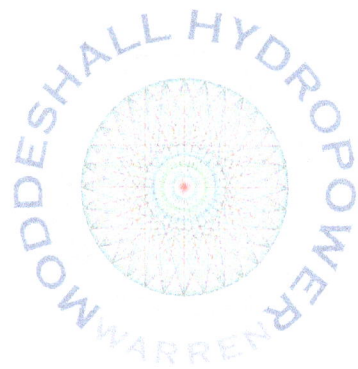

# PART II
## SOME HYDROMETRIC FEATURES OF
## THE MODDESHALL VALLEY SYSTEM

Spot Discharge Assessment at Point A

At around 1135BST I arrived at Point A ( SJ921366 ) and perceived that a brisk brook followed a clean and almost rectilinear trajectory beside the lane. This is the Moddeshall Brook.

The Northern part of the Moddeshall Valley is underlain by a net of several orthogonal faults. The bedrock is vertically displaced in blocks of roughly a kilometer square. Most of the subsoil rock hereabouts is sandstone of Lower Permo-Triassic age but Point A is approximately in the middle of a block of Upper Carboniferous Keele Series sandstones and clays raised as a small horst. This block presents a differential competency with regard to the Permo-Triassic facies and where the Moddeshall Brook enters in the East, at Boar Mill, and where it exists in the West, at the Scotch and Moddeshall Brook confluence, there are highly energetic nick points. The nick point at the former Boar Mill site has been intensified with an artificial millpond berm.

For about a hundred meters the brook's slope appeared to be constant and its unvegetated bed a more or less uniform imbrication of five-centimeter lenticular arenite cobbles.

It seemed to me that this was a straightened and formerly tended mill leat that was gradually reverting to a rivulet's state of nature.

Fifty meters behind me some tall broadleaved trees had shed dry twigs upon the carriageway where vehicles had broken them as they steered to and from a passing place. I removed a selection of these twigs and their fragments to a convenient section, which I marked with a longer twig. Unrolling my Rabone-Chesterman steel rule to its full extent of five meters I used another two-foot twig to mark an upstream section.

Returning to the first section I dunked a third long twig several times across the flow and measured the wetted part as 14 centimeters, having estimated that the leat bed was not more or less than a centimeter variant to that length. Additionally, I extended the rule to measure the width of both sections.

Thirty-five years ago I would have sought to enhance the precision by jumping into the water and grovelling rat-happy within it, but as a ninety-kilo 57-year-old with a gammy knee and a pair of AirWears I felt I better not.

During my half hour at Point A I was happily spared the necessity to explain or justify my unscientific proceedings to bailiffs or passers-by. Only one vehicle passed in this time: A heavy Daimler or Jaguar whose owner paused shyly a hundred meters further up the hill to inspect me in his mirrors.

Breaking where necessary a store of twigs into five-centimeter billets I cast them into mid-stream at the upper section, and used a stopwatch to time their passage to the lower section. All the billets floated proudly, but several disappeared, and I successfully collected times for ten.

These various measurements, and their simple results, are summarised in Table One:-

| | | | | |
|---|---|---|---|---|
| **Stream:** | | Moddershall Brook | | |
| **Grid Reference:** | | SJ921366 | | |
| **Latitude (N):** | Degrees | | 52 | 52.9269 |
| | Minutes | | 55 | |
| | Seconds | | 37 | |
| **Longitude (W):** | Degrees | | 2 | -2.11845 |
| | Minutes | | 7 | |
| | Seconds | | 6 | |
| **Elevation ( meters ):** | | | 139.6 | |
| **Time:** | | 1145BST | | |
| **Date:** | | Thursday 10 September 2009 | | |

**Country Rock:** Carboniferous Keele Series facies
block faulted into PermoTriassic Sandstone

**Weather:** Sunny with slight breeze
No significant rain for seven days
Previous month ( August ) above average rainfall

**Chronometer:** Casio F-91W

**Bottom Condition:** 5cm diameter lenticular cobbles
**Bank Condition:** Feral grassed earth
**Flow condition:** One third bank full

**Section Geometry:** Rectanglular

Measured
Five
Meters

| | | |
|---|---|---|
| **Downstream Section Width ( meters )** | | 1.036 |
| **Upstream Section Width ( meters )** | | 0.91 |
| **Downstream Section Depth ( meters )** | | 0.14 |
| **Downstream Section Depth Plus or Minus Variation ( meters )** | | 0.01 |

**Velocity Tests ( seconds and decimal seconds )**

| | | |
|---|---|---|
| | 1 | 5.11 |
| | 2 | 5.18 |
| | 3 | 4.59 |
| | 4 | 4.87 |
| | 5 | 6.12 |
| | 6 | 4.90 |
| | 7 | 4.98 |
| | 8 | 5.07 |
| | 9 | 4.63 |
| | 10 | 4.85 |

| | |
|---|---|
| **Mean Elapsed Time (s)** | 5.03 |
| **Mean Surface Velocity (m/s)** | 1.006 |
| **Estimated Mean Velocity of Flow (m/s)** | 0.653 |
| **Flow Cross-sectional Area (m$^2$)** | 0.145 |
| **Estimated Discharge ( cumecs )** | 0.0947 |

**Table One**
**Measurement of Leat Flow at Point A**

MODRMILL
Moddeshall Hydropower
    Page 22 of 126
James R Warren
    19:03 Sunday, 19 June 2022

The stream Surface Velocity, $V_s$, of the leat is given by:-

$$V_s = \frac{\sum_{i=1}^{N_T} T_i}{N_T} \times \frac{1}{L}$$

**Equation 1**

where $T_i$ is an individual Transit Time for a surface float, $N_T$ is the Number of Passages logged, and L is the Measured Length of Reach.

The Surface Velocity is not, however, representative of the water velocity over the entire front, which is definable in terms of the Mean Stream Velocity, $V_m$.

The relationship between $V_s$ and $V_m$ is potentially very complex, because the velocity in any given hydrodynamic streamline varies markedly under the influence of boundary friction ( bed and bank "roughness" or rugosity ) as well as inertial effects.

In 1964 the British Standards Institute published an empirical table of Correction Coefficients, $K_c$, against Stream Depth, d, for steady flows in straight canals and other well-behaved streams[2].

In terms of SI depths, linear regression completely epitomises this table for d<1.83 meters as:-

$$K_c = 0.64 + 0.065638d$$

**Equation 2**

Accordingly, an Estimated Mean Velocity may be stated as:-

$$V_m = K_c V_s$$

**Equation 3**

whilst Discharge at the ( Lower ) Section may be declared as:-

$$Q = AV_m$$

**Equation 4**

where A is the water Cross-Sectional Area.

Equation Four is the fundamental hydraulic law, The Equation of Continuity or The Castelli Equation, which was first generalised in 1630.

Because the lower section at Point A is rectangular, Equation Four may be expanded as:-

$$Q = dwV_m$$

**Equation 5**

where d is the Depth and w is Stream Width.

For the data of Point A at the relevant time, $K_c$ was estimated to be 0.649 or $V_m$ to be 64.9% of $V_s$.

The Discharge Q of the Moddeshall Brook computed to be 0.0947 cumecs ( cubic meters per second ).

Together with water pressure, or its surrogate differential head, discharge is the key determinant of hydraulic power.

This discharge value at Point A near to the head of the Moddeshall Valley System may be regarded as setting a lower limit to the emergent potential of the catchment as a power source, and an immediate determinant of the capability of Splashy Mill.

Slope and Rugosity Issues about Point A

The slope and rugosity of the stream are even more problematical than the discharge.

I picked a straight, fast and relatively uniform reach, which occupied a flattish flank between two acclivities of the hillside, but made no attempt to quantify any slope on site.

This flattish section is roughly 93 meters long and is really two straights broken in the fortieth meter by a sinistral 20° crick. The lower subreach of 39.4 meters has a slope of 0.01547619 ( dz/dx ), and the upper, of 53.5 meters length, has a slope of 0.03991228.

The seven-figure accuracies are of course specious as the gradients were assessed post-survey using aerial photographs on MicroSoft VirtualEarth.

Since Point A is in the lower reach I shall discount the larger slope entirely and assume the slope at Point A is 0.01547619.

Having an estimate of Q we may employ Manning's Equation to estimate a Manning Rugosity Coefficient, n, of local friction for the boundary at Point A[3]:-

$$Q = \frac{1}{n} \cdot \frac{A^{\frac{5}{3}} S^{\frac{1}{2}}}{P^{\frac{2}{3}}}$$

**Equation 6**

where Q is Discharge, n is Manning's dimensionless Roughness ( Rugosity ) Coefficient, A is the flow Cross-Sectional Area, S is the Hydraulic Gradient ( taken to be stream surface slope ) and P is the Wetted Perimeter.

Equation Six is the form of Manning's empirical equation that is convenient for open channel flows and may be re-arranged for n as:-

$$n = \frac{1}{Q} \cdot \frac{A^{\frac{5}{3}} S^{\frac{1}{2}}}{P^{\frac{2}{3}}}$$

**Equation 7**

We have our best estimates of Q as 0.0947 cumecs and S as 0.01547619. A is 1.036 meters width times 0.14 meters depth giving 0.145m², whilst:-

$$P = w + 2d$$

**Equation 8**

so that:-

$$P = 1.036 + 2 \times 0.14 = 1.316 \; meters$$

Solving Equation Seven we determine Manning's n to approximate 0.044 at Point A. This is four times as rough as a new wooden launder, but typical for a shallow, meandering river with "noticeable aquatic growth" ( nominally n = 0.045 ) and less than the n = 0.060 said to be typical for a stony stream with weeds and shallows. Both reaches about Point A are clean-bedded with the five-centimeter lenticular cobbles aforementioned, whilst the earthen banks are clad in thick long grass and wild forbs.

A "Standard Natural Stream" has a Manning's Roughness of 0.035, and that may be representative of the Scotch Brook with its silty-sandy bed.

When I saw the leat on 10 September 2009 it was approximately ⅓ bank full. When it is bankfull and weed drag is significant an n=0.06 may reasonably be assumed in default of adequate survey.

Recalculation of Manning's Equation for Q under bankfull conditions and preserved slope indicates a bankfull discharge of 0.341 cumecs, though the actual figure is likely to be slightly more, not only because of less friction than that of full weed drag, but because of increased hydraulic slope.

We may note before leaving the topic of Manning Coefficients that the literature contains several developments of Manning's Equation that yield invalid results when applied to Point A.

At Point A the height of the cobbles proud of the streambed is about one centimeter in a depth of 14 centimeters.

Hydraulic Radius, R, is the quotient of Section Area, A, over Wetted Perimeter, P. In "Geomorphology"[4], the late Richard Chorley and his co-authors advise that when bed particle protrusion is more than a thirtieth of stream depth then the exponent of Hydraulic Radius should be altered. They say to use ¾ rather than ⅔ as the power of R.

This gives a Manning's n of 0.064, a value suited to a choked, low-energy, high-sinuosity watercourse quite unlike the Moddeshall Brook at Point A.

The Point A cobbles have a relatively consistent width on the streambed of about five centimeters.

On Page 12 of a "Guide for Selecting Manning's Roughness Coefficients"[5], the Limerinos Equation is given that purports to yield Manning's n for given values of R and $d_{84}$, where the latter is Particle Diameter, in meters, that equals or exceeds the diameter of 84% of the boundary particles, determined from a random sample of a hundred particles. Furthermore, the Limerinos Equation is claimed to be good for "bed material ranging from small gravel to medium-sized boulders".

The Limerinos Equation employs a logarithm, but the Guide does not make clear whether this is a natural or base-ten logarithm or what. If it is a natural logarithm then the Limerinos Equation predicts a Point A Manning's n of 0.152, an untenable value. Assuming base-ten yields n=0.258, an absurdity. My field-found n of 0.044 gives a $d_{84}$ of 0.000269 meters using a natural logarithm and a $d_{84}$ of $4 \times 10^{-8}$ meters using a base-ten. Both values are absurd: The 14 centimeters deep millrace is not populated with microscopically fine sediment.

Torricelli's Law or The Efflux Law is another basic hydraulic principle general to a number of moving-fluid situations. It was introduced in about 1645 and states:-

$$v = \sqrt{2gH}$$
**Equation 9**

where v is the Stream Velocity, g the local Acceleration due to Gravity, and H the Hydraulic ( Velocity ) Head.

At Point A the local g, uncorrected for altitude or lithology, is around 9.8132764 ms$^{-2}$, whilst our stream velocity was found to approximate 0.653 ms$^{-1}$.

Equation Nine may be rearranged as:-

$$H = \frac{v^2}{2g}$$
**Equation 10**

The Velocity Head therefore computes to be 0.021726128 meters at Point A.

The Estimation of Moddeshall Subcatchments' Areas

A 50000 Ordnance Survey map of The Moddeshall System and its environs was composited from Multimap[5], and I marked the boundary of the total system watershed with a red line. With regard to the marked contours, I divided the map for individual mill incremental catchments with dashed green lines, and delineated the Moddeshall Brook and Cotwalton Brook catchments separately.

I included the system in a 54 square-kilometer reference cartouche upon the blue OS grids.

After improving the resolution of these boundaries with thin black lines I printed the catchment outline map on commercial millimeter graph paper and counted the squares. For the Total Moddeshall System there proved to be 15176 millimeter squares, and for the subcatchments summed individually 15114. Counting error was therefore around 0.4%, which I consider satisfactory to our purpose, given the wide latitudes of error in other regards.

Using The Equation for the Area of an Irregular Polygon, I computed the paper area of the 54Km$^2$ reference cartouche and used it to calibrate the graticule catchment areas for ground area covered.

For the estimation of average subcatchment rainfall I used the isohyet map for the Trent Catchment[6] and overlaid its relevant portion upon the Moddeshall System Catchment Map. The composite map is illustrated below:-

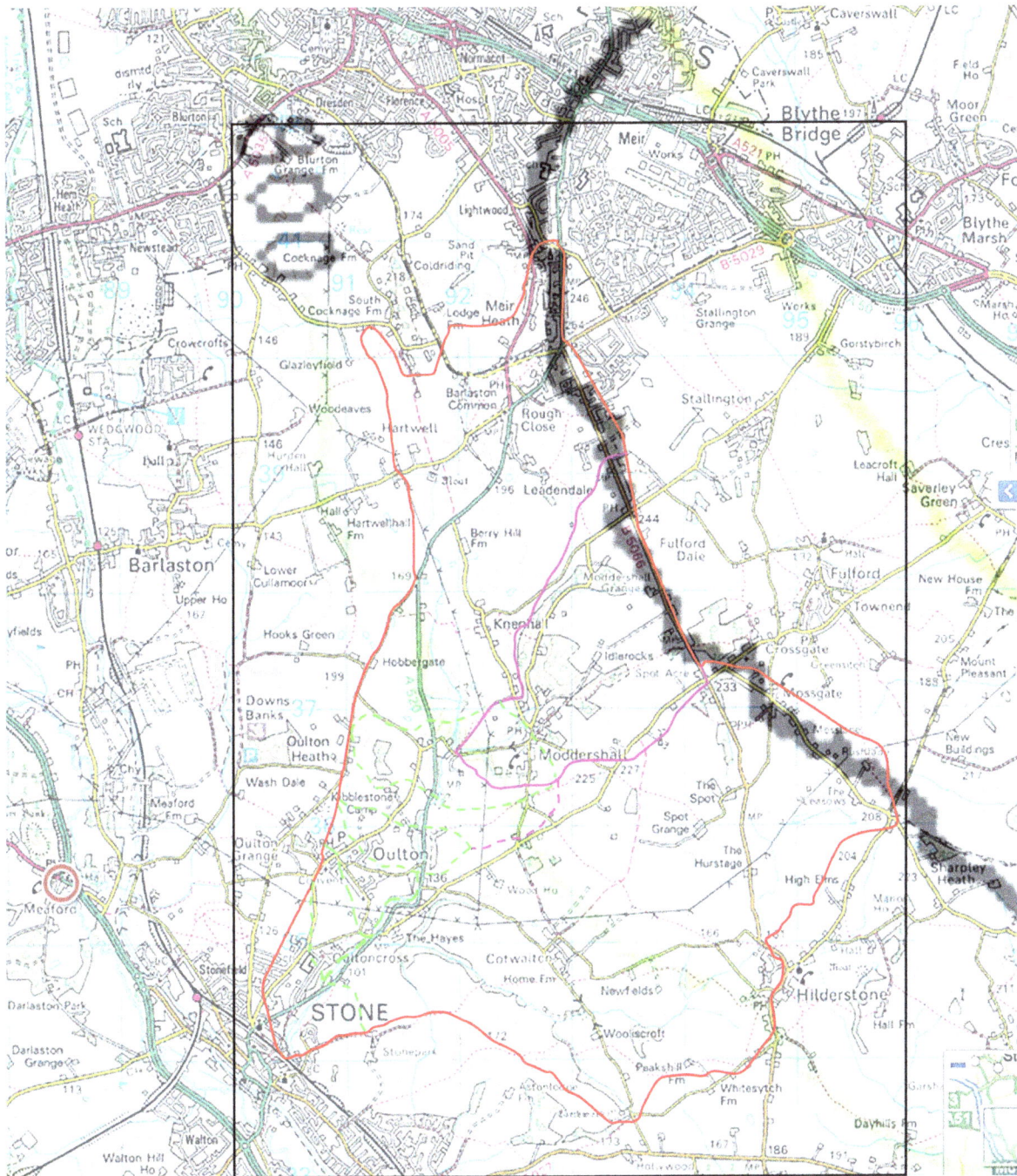

© Crown Copyright 2022 OS 100065726

**Figure One**
**Moddeshall System Catchment Map**

## The Estimation of Centroids and Rainfall

Because I had not tabulated boundary co-ordinates the automatic computation of Moddeshall System centroids was infeasible.

Accordingly, I estimated these by eye.

By placing a ruler normal to isohyets I then interpolated SAAR ( Standard Average Annual Rainfall ) estimates for each subcatchment, working to the nearest millimeter.

It is of course the case that whilst rainfall is traditionally recorded in inches or millimeters it is in fact dimensionally a velocity. This is because it is a length *per* time increment.

To conform with the Systeme International d'Unites, I accordingly reduced (sub)catchment areas to square meters, and rainfall averages to meters per second.

By further recourse to Castelli's Law I was therefore able to compute as discharge the Available Meteoric Precipitation, P, as:-

$$P = AI$$
**Equation 11**

where A is a Catchment or Subcatchment Area, and I is Catchment Average Rainfall.

P is notionally an upper limit to the available water supply before losses, such as evapotranspirative losses, seepage, etc.

Equally, however, P does not include gains due to groundwater emergence.

Neither losses nor gains are publicly available for the Moddeshall System catchments. ( Except, perhaps, at charge ).

Summary estimates of relevant statistics in regard to the nearby Trent Catchment above Darlaston are available in the 2002 UK Surface Water Register[7], and I used these to adjust Moddeshall System P.

At the Darlaston ( SJ885355 ) catchment Mean Annual Rain (P) is given as 842mm of which 616 is Mean Annual Runoff and the balance of 226 Mean Annual Loss. Accordingly, I estimated the Loss Fraction as 0.2684086 and used that to adjust Moddeshall System P to give Apparent Runoff Discharge, $Q_{Runoff}$.

Cumulative Discharge was then tabulated mill-by-mill.

These results are given in Table Two.

From these assessments it is already apparent that the Moddeshall Brook available water estimate is well shy of the spot Point A value that was measured after prolonged dry conditions. This suggests a significant chthonic component of that stream.

| Catchment Code | Catchment Name | Centroid OS Code | Easting | Northing | Method Two 1mm² Square SubTotal | SAAR (mm) | Areas (m²) | SAAR (ms⁻¹) | Apparent Q_runoff Available Precipitation P (m³s⁻¹) | Apparent Q_runoff (m³s⁻¹) | Cumulative Discharge (m³s⁻¹) |
|---|---|---|---|---|---|---|---|---|---|---|---|
| TT | Total Moddeshall System | SJ | 930 | 363 | 15176 | 805.2 | 21332739.13 | 2.5515E-08 | 0.544306624 | 0.398210072 | |
| MB | Moddeshall Brook | SJ | 933 | 373 | 2116 | 812.8 | 2974438.323 | 2.5757E-08 | 0.076613431 | 0.056049731 | |
| CB | Cotwalton Brook | SJ | 939 | 353 | 7135 | 803.0 | 10029592.36 | 2.5447E-08 | 0.255221223 | 0.186717664 | |
| BM | Boar Mill | SJ | 934 | 374 | 1889 | 800.7 | 2655346.877 | 2.5372E-08 | 0.067371532 | 0.049288437 | 0.04928844 |
| SP | Splashy Mill | SJ | 924 | 367 | 277 | 816.2 | 389375.9052 | 2.5864E-08 | 0.010070947 | 0.007367819 | 0.05665626 |
| OM | Ochre Mill ( Scotch Brook Only ) | SJ | 924 | 388 | 3748 | 812.1 | 5268523.079 | 2.5733E-08 | 0.135576376 | 0.099186518 | 0.09918652 |
| ML | Mosty Lee Mill | SJ | 916 | 366 | 496 | 795.3 | 697221.8375 | 2.5201E-08 | 0.01757057 | 0.012854479 | 0.11940882 |
| WM | Wetmore Mill | SJ | 917 | 361 | 429 | 792.1 | 603040.6619 | 2.5099E-08 | 0.015135615 | 0.011073086 | 0.13048190 |
| IM | Ivy Mill | SJ | 915 | 358 | 288 | 788.2 | 404838.4863 | 2.4975E-08 | 0.010110831 | 0.007396998 | 0.13787890 |
| HM | Hayes Mill | SJ | 912 | 355 | 229 | 784.6 | 321902.8242 | 2.4863E-08 | 0.008003457 | 0.005855261 | 0.14373416 |
| CM | Coppice Mill | SJ | 909 | 351 | 233 | 780.0 | 327525.5809 | 2.4717E-08 | 0.008095354 | 0.005922492 | 0.33045183 |
| SM | Stone Mill | SJ | 906 | 345 | 440 | 772.7 | 618503.243 | 2.4485E-08 | 0.015144278 | 0.011079424 | 0.34153125 |
| | Totals Moddeshall System | | | | 15176 | | 21332739.13 | | | | |
| | Totals Component Catchments | | | | 15114 | | 21245586.4 | | | | |
| | Difference | | | | 62 | | 87152.72969 | | | | |
| | Specific Defect | | | | 0.0040854 | | 0.004085398 | | | | |
| | Total | | | | | | 21315870.9 | 2.5176E-07 | 0.2870790 | 0.2100245 | 1.40861806 |
| | Mean | | | | | 796.9 | 2131587.1 | 2.5176E-08 | 0.0318977 | 0.0233361 | 0.15651312 |
| | Population SD | | | | | 13.2 | 3028580.8 | 4.1582E-10 | 0.0406382 | 0.0297305 | 0.10099842 |

**Table Two**
**System Area-SAAR Statistics**

## Comparisons with Local Gauged Statistics

None of the Scotch Brook, Moddeshall Brook or Cotwalton Brook is gauged, but summary statistics for twenty local gauged streams are publicly available[8].

By using the Stream Gauging Station Summary Sheets from the Internet I was able to produce a tabulation, where necessary re-casting measurements in SI terms.

This tabulation is given in Table Three.

It was of interest to determine what light the average character of local streams might shed upon the actual discharge relations within the Moddeshall System.

I decided to study this issue for all twenty local gauging stations ( "All Catchments" ) and for the sixteen isolated largest catchments ( "Independent Catchments" ).

For both sets of gauging stations geographical centroids were computed from OS Grid References. For both sets the centroids proved to be just North-West or North of Werrington, Stoke-on-Trent, at points approximately 13 kms or 8 miles North of Splashy Mill. The OS Grid Reference for All Catchments is SJ927483 and for Independent Catchments only, the centroid grid reference is SJ939496.

By using multiple regression upon the natural logarithms of Discharges, Catchment Areas and Rainfalls I developed empirical equations of the form:-

$$Q = K_F A^{X_A} I^{X_I}$$
**Equation 12**

where Q is a selected Discharge average, $K_F$ is a arbitrary Coefficient, and $X_A$ and $X_I$ are respective Exponents of Area and Rainfall. This is the Free Exponent Fitment Form.

It is clear that dimensional homogeneity is preserved when $X_A = X_I = 1$. Therefore, I also studied the quality of fitments for a family of equations defined by:-

$$Q = K_X A^1 I^1 = K_X P$$
**Equation 13**

where P is the apparent Available Precipitation ( in cubic meters per second ). This is the Fixed Exponent Fitment Form.

The K and X coefficients are tabulated in Table Four, together with the fitment quality metric, $R^2$, the Regression Coefficient of Determination.

| Serial | River | Gauging Station | Station Number | OS Code | Easting | Northing | Catchment Area (m²) | Level of Station (mOD) | Max Altitude (mOD) | Mean Flow (m³s⁻¹) | Q95 Flow (m³s⁻¹) | Q10 Flow (m³s⁻¹) | 61-90 Annual Average Rainfall (ms⁻¹) |
|---|---|---|---|---|---|---|---|---|---|---|---|---|---|
| 1 | Bollin | Dunham Massey | 69006 | 33 SJ | 727 | 875 | 256000000 | 12.8 | 483 | 4.45 | 1.209 | 9.073 | 2.79172E-08 |
| 2 | Burbage Brook | Burbage | 28070 | 43 SK | 259 | 804 | 9100000 | 290 | 451 | 0.17 | 0.025 | 0.36 | 3.18148E-08 |
| 3 | Churnet | Basford Bridge | 28061 | 33 SJ | 983 | 520 | 139000000 | 132.9 | 451 | 1.89 | 0.431 | 4.191 | 3.09276E-08 |
| 4 | Coley Brook | Coley Mill | 54099 | 33 SJ | 779 | 192 | 37300000 | 68.8 | 124 | 0.32 | 0.12 | 0.59 | 2.22133E-08 |
| 5 | Dane | Hulme Walfield | 68006 | 33 SJ | 845 | 644 | 150000000 | 66.4 | 547 | 2.39 | 0.411 | 5.38 | 3.22585E-08 |
| 6 | Dane | Rudheath | 68003 | 33 SJ | 668 | 718 | 407100000 | 13.2 | 547 | 4.9 | 0.934 | 10.62 | 2.70616E-08 |
| 7 | Dove | Rocester Weir | 28008 | 43 SK | 112 | 397 | 399000000 | 86.3 | 550 | 7.5 | 1.739 | 15.67 | 3.23535E-08 |
| 8 | Goyt | Marple Bridge | 69017 | 33 SJ | 964 | 898 | 183000000 | 74.4 | 547 | 3.7 | 0.767 | 8.179 | 3.65047E-08 |
| 9 | Hamps | Waterhouses | 28041 | 43 SK | 82 | 502 | 35130000 | 210.2 | 489 | 0.7 | 0.061 | 1.66 | 2.83292E-08 |
| 10 | Henmore Brook | Ashbourne | 28058 | 43 SK | 176 | 463 | 42000000 | 116.4 | 379 | 0.45 | 0.075 | 1.004 | 2.4178E-08 |
| 11 | Meece Brook | Shallowford | 28079 | 33 SJ | 874 | 291 | 86300000 | 81.1 | 162 | 0.65 | 0.142 | 1.251 | 2.20866E-08 |
| 12 | Penk | Penkridge | 28053 | 33 SJ | 923 | 144 | 272000000 | 76.3 | 151 | 2.25 | 0.594 | 4.159 | 2.38294E-08 |
| 13 | Sow | Great Bridgeford | 28052 | 33 SJ | 883 | 270 | 163000000 | 77.1 | 168 | 1.19 | 0.351 | 2.261 | 2.26253E-08 |
| 14 | Sow | Milford | 28014 | 33 SJ | 975 | 215 | 591000000 | 69.4 | 195 | 6.17 | 1.5 | 12.02 | 2.59525E-08 |
| 15 | Trent | Darlaston | 28083 | 33 SJ | 885 | 355 | 195200000 | 85.5 | 277 | 3.8 | 1.508 | 6.878 | 2.52554E-08 |
| 16 | Trent | Great Heywood | 28006 | 33 SJ | 994 | 231 | 325000000 | 69.5 | 330 | 4.45 | 2.144 | 7.334 | 2.73468E-08 |
| 17 | Trent | Stoke-on-Trent | 28040 | 33 SJ | 892 | 467 | 53200000 | 113.2 | 213 | 0.65 | 0.128 | 1.376 | 2.27837E-08 |
| 18 | Weaver | Audlem | 68005 | 33 SJ | 653 | 431 | 207000000 | 44.7 | 222 | 1.61 | 0.228 | 3.908 | 2.32591E-08 |
| 19 | Wistaston Brook | Marshfield Bridge | 68004 | 33 SJ | 674 | 552 | 92700000 | 30.1 | 221 | 0.93 | 0.222 | 1.798 |  |
| 20 | Wye | Ashford | 28023 | 43 SK | 182 | 696 | 154000000 | 139 | 549 | 3.24 | 0.979 | 6.116 | 3.69483E-08 |

**Table Three**
**Selected Parameters of West Mercian River Gaugings**

**FREE**

| Catchments | Discharge Statistic | Coefficient $K_F$ | Exponent of A $x_A$ | Exponent of I $x_I$ | $R^2$ |
|---|---|---|---|---|---|
| All | $Q_{mean}$ | 1100848.59 | 1.03982377 | 1.88563865 | 0.96710752 |
| All | $Q_{95}$ | 0.12920326 | 1.17455211 | 1.19887227 | 0.87579694 |
| Independent | $Q_{mean}$ | 1103889.48 | 1.03284404 | 1.87871657 | 0.98229938 |
| Independent | $Q_{95}$ | 1.8196946 | 1.16124995 | 1.33838115 | 0.90167424 |

**FIXED**

| Catchments | Discharge Statistic | Coefficient $K_X$ | Exponent of P $x_P$ | $R^2$ |
|---|---|---|---|---|
| All | $Q_{mean}$ | 0.44654078 | 1.03480283 | 0.94898371 |
| All | $Q_{95}$ | 0.08462228 | 1.17440774 | 0.87578620 |
| Independent | $Q_{mean}$ | 0.44928354 | 1.02652659 | 0.96432091 |
| Independent | $Q_{95}$ | 0.08352303 | 1.15992704 | 0.90109726 |

**Table Four**
**Empirical Discharge Equations for West Mercian Catchments**

$R^2$ expresses the degree of variation in the dependent variable that is accounted for by the regressed independent variables, in this case A and I. The variation accounted for is expressed as a fraction of the total.

On this basis it can be seen that Free fittings are as good or superior to Fixed, and Independent Catchments better than promiscuous mixes.

The best equation with $R^2$ = 0.98229938 is for Mean Discharge, $Q_{mean}$, versus catchment A and I for the sixteen Independent Catchments:-

$$Q_{mean} = 1103889.48 A^{1.03284404} I^{1.87871657}$$

**Equation 14**

The best equation, with $R^2$ = 0.90167424, for Discharge Exceeded Ninety-Five Percent of the Time, $Q_{95}$, is:-

$$Q_{95} = 1.8196946 A^{1.16124995} I^{1.33838115}$$

**Equation 15**

It makes more sense to base power availability estimates upon a water supply that is provided on nineteen occasions out of twenty rather than roughly half the time, even for Grid connected facilities. Therefore it is unfortunate that our estimate of $Q_{95}$ is less reliable than that of $Q_{mean}$.

On a more positive note, it is clear that the agreement between actual and predicted discharges is markedly less erratic for small streams than for large, since the brooks of The Moddeshall System are modest indeed.

That concordance is illustrated in Figure One:-

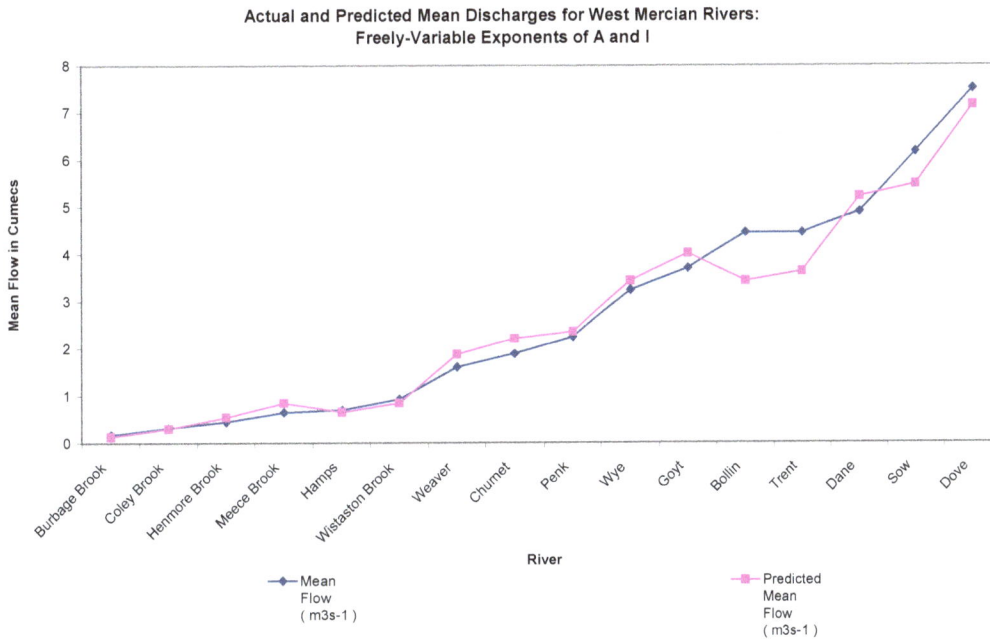

Actual and Predicted Mean Discharges for West Mercian Rivers:
Freely-Variable Exponents of A and I

**Figure One**
**Actual and Predicted Mean Discharges at**
**West Mercian Independent Gauging Stations**

When, however, the local empirical estimates of $Q_{mean}$ and $Q_{95}$ are transferred to the AI context of The Moddeshall System many absurdities appear, as can be seen from Table Five.

Calibrated discharge values not merely exceed P, but are those appropriate to the tidal sections of major British rivers rather than little millstreams.

But on the other hand, indicated Mean Discharge at Splashy Mill is fully 25.9-times less than at Point A just upstream, whilst $Q_{95}$ at Splashy Mill is a mere 395ml per second, less than a pint and manifestly inadequate to a wheel.

There is really no substitute for the gauging of this catchment, especially in the known presence of major sustained abstractions. The nearby Cresswell complex of water utility draughts from the Permo-Triassic country rock takes 29.8 megaliters or 29800 tonnes per day[9] ( this is equivalent to 0.0009443 $m^3s^{-1}$ ), and there are further subtractions and additions for both domestic supply and fish farm use.

As a minimum, the Moddeshall and Cotwalton Brooks ought to be gauged at their respective confluences with the Scotch Brook, and the Scotch Brook itself above its confluences with them.

| Catchment Code | Catchment Name | Areas ($m^2$) | SAAR ($ms^{-1}$) | Assessed $Q_{95}$ ($m^3s^{-1}$) | Assessed $Q_{mean}$ ($m^3s^{-1}$) | Measured Q ($m^3s^{-1}$) | $Q_{95}$ Calibrated against Point A | $Q_{mean}$ Calibrated against Point A |
|---|---|---|---|---|---|---|---|---|
| | | | | Apparent $Q_{95}$ | Apparent $Q_{mean}$ | | Calibrations | |
| TT | Total Moddeshall System | 213327739.13 | 2.5515E-08 | 0.04057456 | 0.22243925 | | 9.71838837 | 5.76962366 |
| MB | Moddeshall Brook | 2974438.32 | 2.5757E-08 | 0.00416995 | 0.02959207 | | 0.9987434 | 0.76755832 |
| CB | Cotwalton Brook | 10029592.36 | 2.5447E-08 | 0.01682990 | 0.10150727 | | 4.03108438 | 2.63289294 |
| BM | Boar Mill | 2655346.88 | 2.5372E-08 | 0.00358213 | 0.02558451 | | 0.85758878 | 0.66361044 |
| SP | Splashy Mill | 389375.91 | 2.5864E-08 | 0.00039547 | 0.00365192 | 0.09472338 | 0.09472338 | 0.09472338 |
| OM | Ochre Mill ( Scotch Brook Only ) | 5268523.08 | 2.5733E-08 | 0.00808927 | 0.05331548 | | 1.93753531 | 1.38289569 |
| ML | Mosty Lee Mill | 697221.84 | 2.5201E-08 | 0.00075129 | 0.00634788 | | 0.1799493 | 0.16465116 |
| WM | Wetmore Mill | 603040.66 | 2.5099E-08 | 0.00063134 | 0.00542282 | | 0.15121875 | 0.14065696 |
| IM | Ivy Mill | 404838.49 | 2.4975E-08 | 0.00039484 | 0.00355992 | | 0.09457117 | 0.09233706 |
| HM | Hayes Mill | 321902.82 | 2.4863E-08 | 0.00030074 | 0.00278577 | | 0.07203365 | 0.07225727 |
| CM | Coppice Mill | 327525.58 | 2.4717E-08 | 0.00030444 | 0.00280478 | | 0.07291876 | 0.07275038 |
| SM | Stone Mill | 618503.24 | 2.4485E-08 | 0.00062900 | 0.00531363 | | 0.15065746 | 0.1378248?2 |
| | Totals Moddeshall System | | | | | | | |
| | Totals Component Catchments | | | | | | | |
| | Difference | | | | | | | |
| | Specific Defect | | | | | | | |
| | Total | 21315870.9 | 2.5176E-07 | 1.81969460 | 1103889.48 Coeff | | Local Independent Catchments see | |
| | Mean | 2131587.1 | 2.5176E-08 | 1.16124995 | 1.03284404 Exp for Area | | MODQIREG.xls | |
| | Population SD | 3028580.8 | 4.1582E-10 | 1.33838115 | 1.87871657 Exp for SAAR | | IND REG FREE Q95 | |
| | | | | 0.90167424 | 0.98229938 $R^2$ | | | |

**Table Five**
**Untenable Moddeshall System Discharges indicated by**
**West Mercian River Gauging**

## Stream Slope Topography and Discharge

I wondered whether the seemingly precise aerial photographs of Google Earth or MicroSoft Virtual Earth would furnish good co-ordinates and elevations, sufficient to the accurate estimation of stream slopes.

Google gave terrestrial co-ordinates to a hundredth of a second of arc ( about 0.3086 ground meters at Ivy Mill ), but elevations only to the nearest meter. Virtual Earth gave up to four decimals of a degree ( ± 11.11 meters ) but heights to the nearest foot.

Initially I attempted to use VE to assess the differential heights between heads and tails at mills but at several locations VE gave higher tailraces than headraces. At other wheel-pits, the apparent drop was manifestly too small for the wheels that were historically documented or that I had seen. Or indeed for *any* useful roman wheel. When I appealed to Google Earth it agreed with VE entirely, only in Metric.

It was thus that my naive, yet suspicious, belief in their specious precision was disabused, and when it dawned upon my slow appreciation that most of the watercourse was hidden beneath summer trees I realised that I must find a better mapping.

I remembered that many years ago a man at Glasgow University had counselled me not to use unadjusted aerial photographs for geodesy, advice that I had taken.

So I downloaded 1:25000 OS map images of the Scotch and Moddeshall Brook catchments piecemeal via Multimap and butted them together with my old MS PhotoEdit program. I grouped the collected objects and saved the resulting image as a jpeg. I did this also with the Cotwalton Brook basin, but I decided not to proceed with analysis of that stream for three reasons: The Cotwalton Brook has significant sinuosity; it has but slight history of hydropower exploitation; and its final submission to the Scotch Brook is known in approximation.

I loaded the Scotch-Moddeshall jpeg back into PhotoEdit and kidded the software that I intended to draft a line. In the upper part of the image I read the PhotoEdit pixel co-ordinates of the corners of a block of blue OS kilometer-square graticule lines. The bounding longitude lines were two kilometers separate and the two latitude lines four. The mean errors in the oblation of the square turned out to be $432.75/434 \equiv 0.28802\%$ ( pixels ) and the average calibration scaler was therefore 2.307466 meters/pixel.

It seemed best to log Cartesian pixel co-ordinates at every intersection of an orange 5-meter OS contour with the stream course in order to build a topographic profile of the river.

Our motive is of course to establish the hydraulic energy slope of the Scotch-Moddeshall stream, a function of the inclinations of the water and streambed surfaces. At our scale of examination, however, both those slopes are approximated well within the limits of other error by the topographical stream slope.

Arguably the least of these extraneous errors was the decision to simplify the horizontal course delineation by eschewing all curvilinear approximations in favor of a simple Pythagorean link between each contour intersection. This accordingly defined the course as a series of straight-line segments defined by:-

$$d_i = \sqrt{(x_i - x_{i-1})^2 + (y_i - y_{i-1})^2}$$

**Equation 16**

where $d_i$ is the ith Stream Segment Length, $x_i$ is the ith Horizontal Co-ordinate and $y_i$ is the ith Vertical Co-ordinate. Both x and y are scaled to meters.

Accordingly, the Cumulate Stream Course Length, $C_i$, is given by:-

$$C_i = \sum_{j=0}^{i} d_j$$

**Equation 17**

The datum for the Course Length is the intersection with the 85-meter contour at a point near to Stone Mill.

In addition to these points, subsidiary 3D co-ordinates were taken at the lower boundaries of the defined mill catchment areas, at positions where their lades used to abstract water. This enabled cumulative runoff discharges to be associated with particular 5-meter intervals immediately downstream.

The resulting assessments are summarised in Table Six.

Figure Two gives an impression of the stream course plan as modelled by the recorded straight-line segments:-

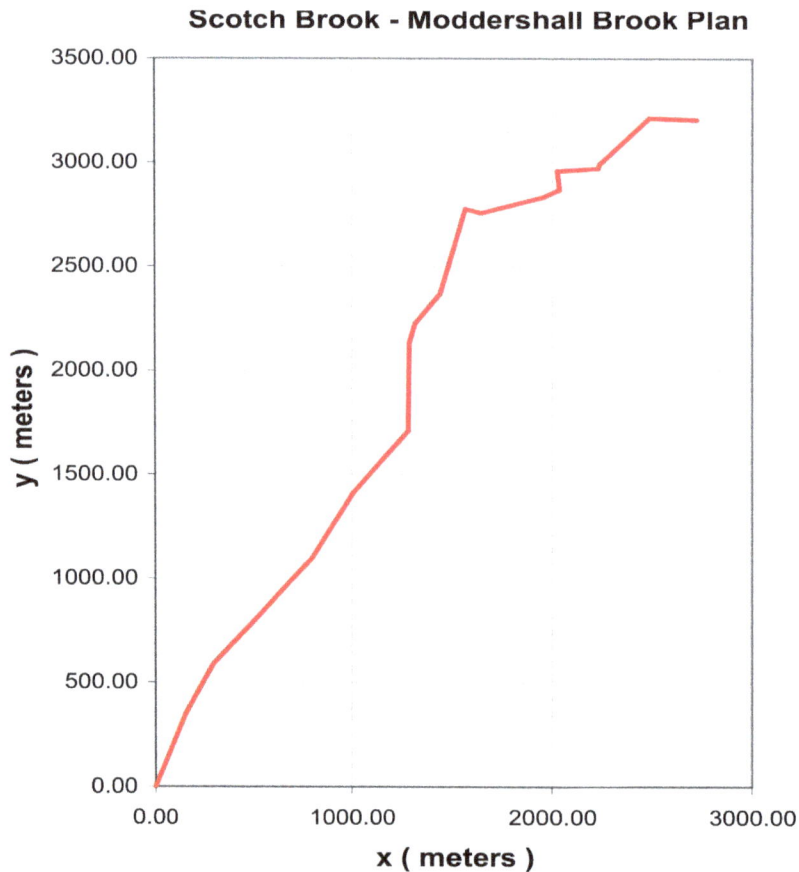

**Figure Two**
**Plan of Scotch-Moddeshall Stream**

**RIVER:** Scotch Brook/Moddeshall Brook

| Serial | Contour | Course Distance $C=\Sigma d$ | $_3S$ | $d_{3S}/dC$ | Deterministic Discharge Assumption (cumecs) Apparent $Q_{Runoff}$ ($m^3 s^{-1}$) | $\Sigma Q$ on Contour | Stream Power Density ($Wm^{-1}$) | Reach Power ( Watts ) | Notes |
|---|---|---|---|---|---|---|---|---|---|
| 1 | 80 | 0.00 | 0.006490 | 0.00002307 | 0.011079 | 0.210025 | 13.372248 | 8913.271755 | |
| 2 | 85 | 666.55 | 0.012399 | -0.00000307 | 0.005922 | 0.198945 | 24.198445 | 26243.923430 | Stone Mill Take-off |
| 4 | 95 | 1751.08 | 0.012826 | -0.00000199 | 0.005855 | 0.193023 | 24.286424 | 9911.654450 | |
| 6 | 105 | 2159.19 | 0.012014 | 0.00004667 | 0.007397 | 0.187167 | 22.057880 | 9365.750198 | |
| 7 | 110 | 2583.79 | 0.031832 | 0.00007803 | 0.011073 | 0.179770 | 56.135235 | 5409.338021 | |
| 8 | 115 | 2680.16 | 0.039351 | -0.00010790 | 0.012854 | 0.168697 | 65.121600 | 12142.747991 | |
| 9 | 120 | 2866.62 | 0.019232 | 0.00003679 | 0.099187 | 0.113951 | 21.498381 | 11068.201931 | |
| 10 | 125 | 3381.46 | 0.037048 | -0.00000328 | 0.007368 | 0.056656 | 20.590611 | 14946.054167 | Scotch Brook Headwater |
| 12 | 135 | 4107.32 | 0.121741 | -0.00031321 | 0.049288 | 0.049288 | 58.863009 | 19525.200224 | |
| 17 | 160 | 4439.03 | 0.017848 | -0.00007361 | | | | | |
| 18 | 165 | 4681.49 | | | | | | | |
| 19 | 170 | | | | | | | | |
| **Totals** | | 29316.693961 | 0.310781 | -0.000318 | 0.210025 | 1.357523 | 306.123833 | 117526.142167 | |
| **Means** | | 2665.153996 | 0.031078 | -0.000032 | 0.023336 | 0.150836 | 34.013759 | 13058.460241 | |
| **Pop SDs** | | 1415.550681 | 0.032078 | 0.000107 | 0.029731 | 0.058325 | 18.771535 | 5993.218259 | |

**Table Six**
**Scotch-Moddeshall Stream Profile Summary**

Figure Three represents the streams' topographic profile along their course.

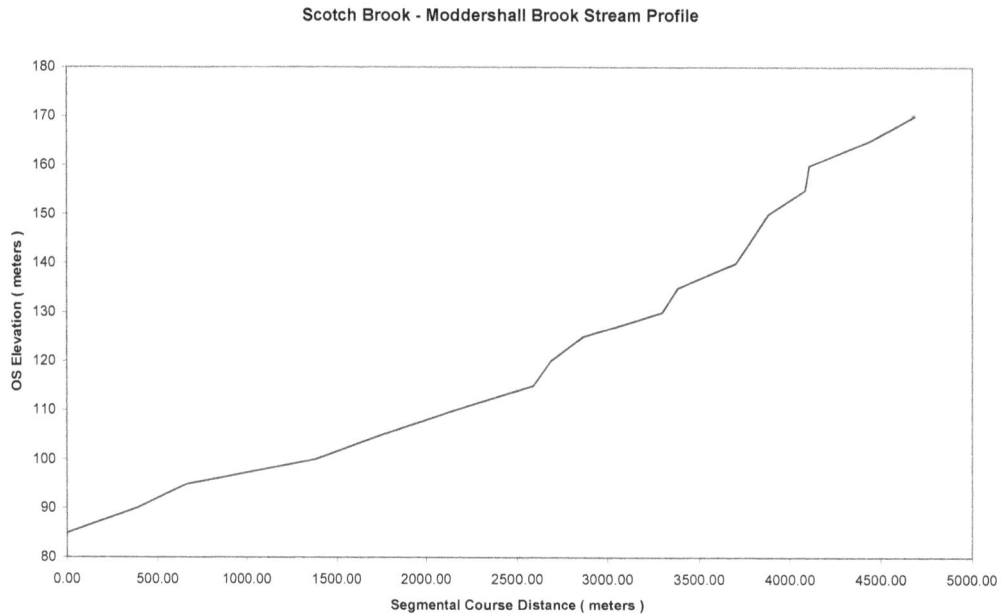

Scotch Brook - Moddershall Brook Stream Profile

**Figure Three**
**Scotch-Moddeshall Brooks' Topographic Profile**

Note that there are five distinct geomorphological nick-points, including a very abrupt one in the higher headwaters of the system. Almost any upstanding feature that crosses the streambed can cause such nick-points. In nature, they are often caused when relatively obdurate sandstone or quartzite strata are over-ridden by tumbling water, and even more spectacular falls arise when streams cross intrusive igneous sills and dykes. Cultural features may also create nick-points, as for example dams and weirs. All these nick-points betray points of potential power concentration. The sharp nick-point near to Meter 4100 is due to the old dam of the Boar Mill millpond, as seen in the photograph "Boar Mill Pond in Moddeshall".

As aforementioned, cumulate discharges were associated with spacial junctures, and these are plotted in Figure Four. The abrupt ascent around the 2700th meter is due to the union of the Moddeshall and Upper Scotch Brooks below Splashy Mill. A more modest rise is apparent near Meter 600, where the Cotwalton Brook joins the Scotch Brook.

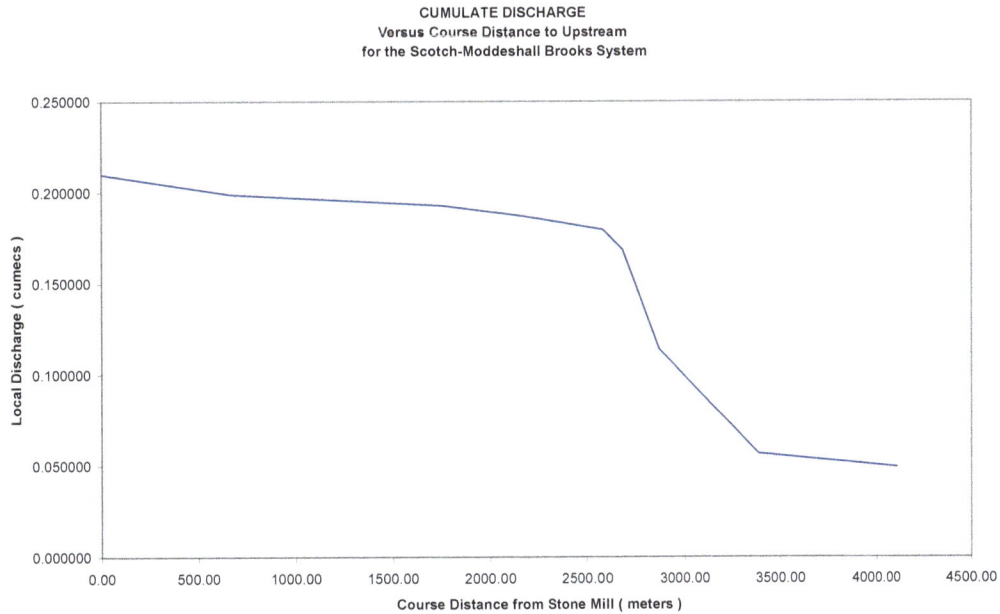

CUMULATE DISCHARGE
Versus Course Distance to Upstream
for the Scotch-Moddeshall Brooks System

**Figure Four**
**Cumulate Discharge along the Watercourse**

Stream Slope

Two estimates of topographical stream slope were computed from the assembled data.

"One-Point" Slope, $_1s_i$, was computed as the simple rectilinear back gradient to Point i:-

$$_1S_i = \frac{z_i - z_{i-1}}{d_i - d_{i-1}}$$

**Equation 18**

where $z_i$ is the Elevation at Point i, and additionally a "Three-Point" Average Slope at Point i, defined in terms of $_1s_i$, by:-

$$_3S_i = \frac{_1S_{i+1} - _1S_i}{2}$$

**Equation 19**

In theory, $_3s_i$ gives a more accurate estimate of slope than $_1s_i$.

Linear regressions of the two resulting data strips of dz/dC against C produced nearly identical full-course average gradients of about $1.53 \times 10^{-5}$, though $_1s$ variations were of double the $_3s$ amplitudes especially in the upper reaches. Quadratic regression of $_3s$ produced a minimum slope at about Meter 1030.

These course slope findings are illustrated below:-

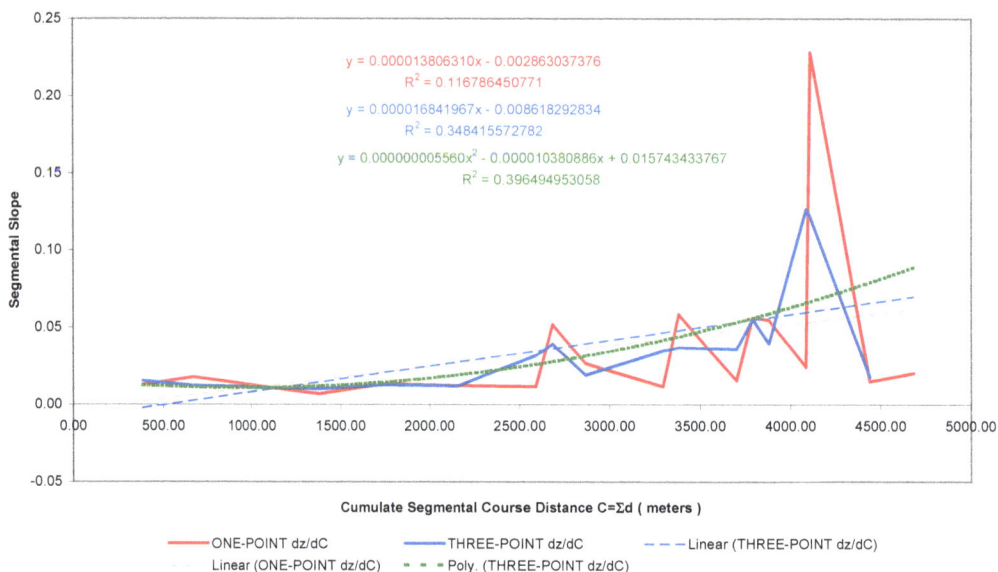

Scotch Brook - Moddeshall Brook Stream Slope Versus Distance

**Figure Five**
**Stream Slope Variation**

The Lineal Change of Stream Slope was computed for $_3s$ values. This forward function is defined by:-

$$\frac{d_3s}{dC} = \frac{_3s_{i+1} - _3s_i}{C_{i+1} - C_i}$$

**Equation 20**

Lineal Change of Slope was erratic in the upper half of the catchment but regressed to zero within limits of error, the actual statistical zero occurring near to C=2450 meters.

Therefore, at least for the Scotch-Moddeshall Brook system we may write:-

$$\frac{d_3s}{dC} \approx 0$$

**Equation 21**

The pattern of $d_3s/dC$ behaviour with C is shown in Figure Six.

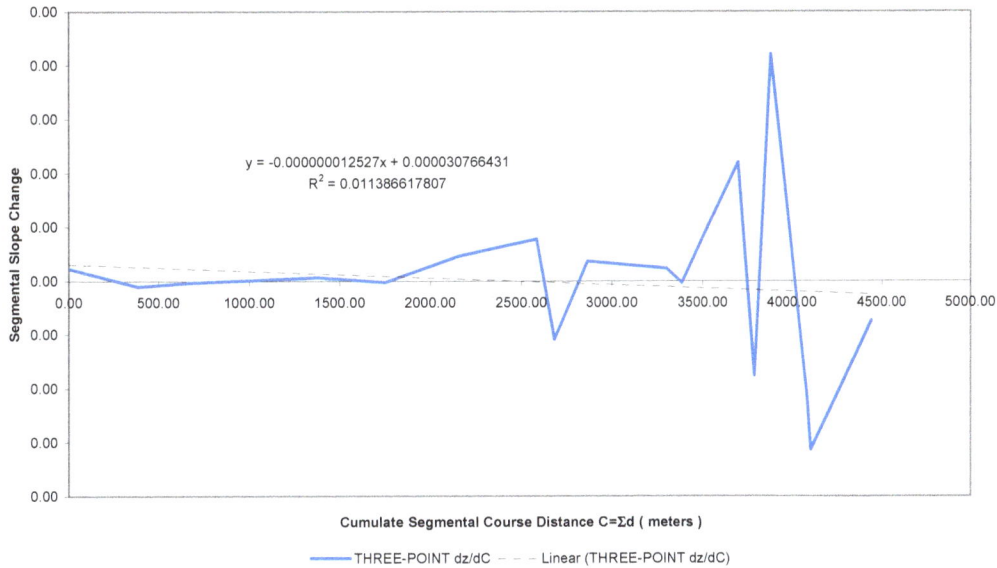

**Figure Six**
**Stream Slope Change with Course Distance**

## Stream Slope Breaks

The four largest stream slope breaks are tabulated below:-

**RIVER:** Scotch Brook/Moddeshall Brook

| Contour | Course Distance C=Σd | Location Description |
|---|---|---|
| 95 | 666.55 | Just above Stone Mill Take-off |
| 115 | 2583.39 | 100m North of Ivy Mill |
| 130 | 3295.83 | Splashy Mill Bridge |
| 160 | 4085.43 | Boar Mill Site |

**Table Seven**
**Important Stream Slope Breaks**

## Lineal Stream Power Density

Stream Power Density, $\Omega$, has the dimensions $MLT^{-3}$. It is often confusingly called "Stream Power", which it is not. Power is energy transacted per unit time and has the dimensions $ML^2T^{-3}$. Therefore it is clear that $\Omega$ is the *lineal density* of stream power per ( say ) meter of the stream course, or in SI terms, Watts per meter.

To assess Stream Power proper you must multiply $\Omega$ by the length of the river or rivulet segment through which it operates.

Stream Power Density, $\Omega$, is given by:-

$$\Omega = \gamma Q s$$

**Equation 22**

where $\gamma$ is the Specific Weight of Water, Q is the Discharge and s is the Stream Energy Slope.

The context of Q and s is local and $\Omega$ varies continually along a river.

The context of $\gamma$ is also chorological, though at a more regional level.

$$\gamma = \rho g$$

**Equation 23**

where $\rho$ is the Density of Water ( kg/m$^3$ ) and g is the Acceleration Due to Terrestrial Gravity ( m/s$^{-2}$ ).

I do not know the average temperature of stream water in The Moddershall System. I adopted the mean temperature of the Severn at Shrewsbury as an approximation, though much of that flow is of montane origin, and Shrewsbury is thirty miles from Stone. The average temperature of the Severn at Shrewsbury is 10.6°C.

The CRC Handbook of Chemistry and Physics[10] specifies a five-degree polynomial equation divided by a single-degree polynomial that computes ( pure ) water density to seven-figure accuracy for any given temperature.

According to that, the density of water at 10.6°C is 999.6445 kg/m$^3$, and that is the value of $\rho$ I adopted for Moddeshall power calculations.

With regard to g the CRC Handbook tabulates guide values at each degree that range from 981.071 at 50°N to 981.507 at 55°N.

Using all six tabulated pairs between those latitudes I found that the following equation epitomises the variation of Gravitational Acceleration with Latitude in English latitudes:-

$$g = 9.7670333333 + 0.0008737143\phi$$

**Equation 24**

where $\phi$ is Latitude.

The latitude of Ivy Mill, in the approximate center of the Scotch-Moddeshall Catchment, is 52.9188611°N, where the Earth's Radius is nominally 6364472.4 meters.

At Ivy Mill the computed g is 9.8132693ms$^{-2}$ and this was adopted as the g for the whole basin.

As the product of $\rho$ and g, Specific Weight $\gamma$ was therefore determined as 9809.7805 Newtons/m$^3$.

In the context of our Catchment, Stream Power Density could be specified as:-

$$\Omega_{Runoff} = \gamma Q_{Runoff} \, {}_3s$$
$$\textbf{Equation 25}$$

where $Q_{Runoff}$ was accumulated at critical contours for an assigned ${}_3s$ value at the point.

It was therefore possible to associate a specific $\Omega_{Runoff}$ with each of the nine critical reaches that influenced historical Catchment mills. Simple multiplication of local $\Omega_{Runoff}$ by the relative reach length then gave the nominal stream power generated within that river segment.

These results are tabulated in the Reach Power Column of Table Six and we may note in particular these three major Power Density peaks:-

**RIVER:** Scotch Brook/Moddeshall Brook

| Contour | Course Distance $C=\Sigma d$ | Stream Power Density ( $Wm^{-1}$ ) | Reach Power ( Watts ) | Location Description |
|---|---|---|---|---|
| 95 | 666.55 | 24.198445 | 26243.923430 | Just Above Stone Mill Take-off |
| 120 | 2680.16 | 65.121600 | 12142.747991 | At Road Bridge 100m N of Wetmore Mill |
| 160 | 4107.32 | 58.863009 | 19525.200224 | Boar Mill Dam |

**Table Eight**
**Important Power Density Peaks**

The plot of Figure Seven is the Power Profile for the Scotch Brook – Moddeshall Brook stream contiguum showing Local Lineal Stream Power Density, $\Omega$ ( W/meter ) against Course Distance, C.

A linear regression of Catchment $\Omega$ versus C provides this estimator ( SI units ):-

$$\Omega = 15.03190167 + 0.00845887C$$
$$\textbf{Equation 26}$$

which has a Coefficient of Determination, $R^2$, of 0.29857018.

The low $R^2$ suggests, correctly as our science determined, that the drastic variations of $\Omega$ are real environmental effects rather than noise.

Otherwise, $\Omega$ increases with elevation as we would expect from the geomorphological doctrine of stream maturity.

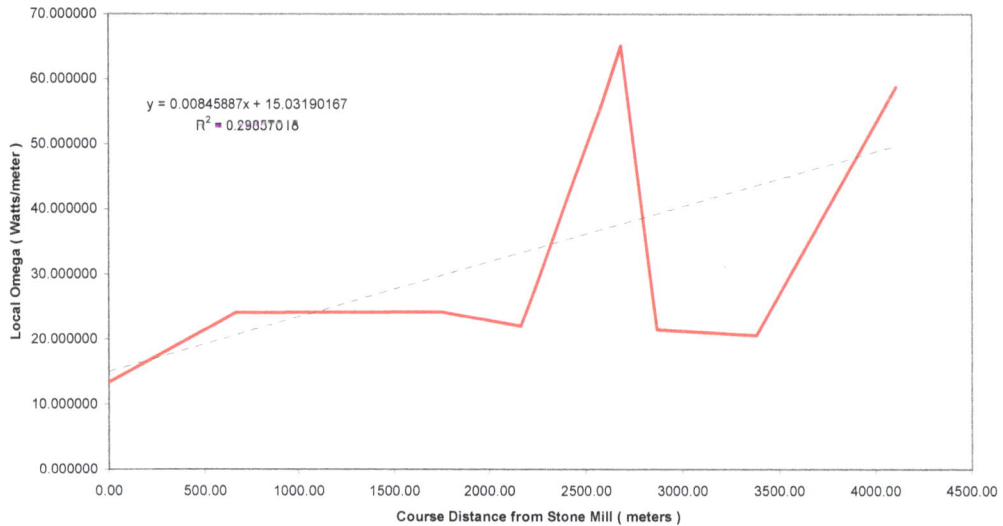

**Figure Seven**
**Scotch-Moddeshall Watercourse Power Density Profile**

Stream Power

Stream Power, W, in Watts, is given by:-

$$W = \sum \Omega_{Runoff}$$
**Equation 27**

in a given section of stream course.

We may note that of the nine mill reaches all but one generate at least 10Kw of stream power and five well over that figure. The high headwaters of the Moddeshall Brook that bore down upon Boar Mill in Moddeshall Village create 20Kw of stream power, and the Cotwalton Confluence reach above Stone Mill take-off contributes 26Kw.

The Romans thought of rivers as gods and goddesses. Our greatest English rivers still bear their names. If you perturb a stream it shall react defensively, like a living creature endued with spirit.

Stream Power is a natural phenomenon. It embodies the energy that a river uses to stir its water ( 97% ); shift sediment around; and generally mind its own business.

The river's purpose is not the electrification of terrestrial apes. Stream Power may or may not be partially abstractable. Any attempt to remove part of it may have serious human consequences, especially in terms of water supply deprivation, waste removal, flooding, erosion and silting.

If stream power delineation has merit for run-of-river hydropower exploitation then that merit is in the indication of potentially-advantageous wheel or turbine sites.

# PART III
## PRINCIPLES OF WATERWHEEL DYNAMICS

The Input Gate Power, G, of a hydraulic machine, G, may be summarised as:-

$$G = \gamma QH$$

**Equation 28**

where H is Hydraulic Head, a composite surrogate of several energetic gain and loss factors.

G may be viewed as the natural power available at the valve or sluice proximal to the wheel or turbine, and does not involve mechanical or electrical losses due to product power transmission or use.

I should emphasis that this Gate Power, G, is a natural phenomenon intrinsic to moving water ( *effet naturale* ). It is not necessarily abstractable, even in part. Any energy removed from this action depends upon the type of transducive machinery employed and its environmental context ( *effet generale* ).

I have read claims that the waterwheel has done more for human welfare than all the rulers, priests, philosophers and artists in history. That may or may not be, but certainly it has achieved more for the emancipation of women than has female suffrage.

Several authors have found the phrase "vertical water wheel" descriptive but tedious, at least in repetition. The phrase "roman wheel" is briefer and places the innovation in an ancient history that predicated the modern world.

The phrase "gravity wheel" is sometimes encountered, and is apt to the overshot, pitchback and breast variants of the roman wheel genus, whilst "impulse wheel" is sometimes used to describe undershot types.

Around 25BC Vitruvius described the use of what was probably an undershot waterwheel and a certain example of such has been detected as a cast in the welded volcanic dust of Pompeii. An overshot waterwheel, dynamically a different entity, was drawn in the Christian Catacombs of Rome, and like the Pompeii wheel that dates to our First Century with Christ[11].

Maybe a century earlier Antipater eulogised the waterwheel in libertarian terms like mine, though better limned. On the whole, however, the Ancients were reluctant to discuss, much less celebrate, machinery. They regarded it as servile, and like all slave societies, their snobbery limited their own advancement.

In our land the Anglo-Saxons, proud and free, brought us the civil waterwheel, and when the Norman subdued the Saxon and counted his assets he found that by 1086 the English had 5624 watermills in a terrain noted for no Niagaras.

The Newtonian dynamics of moving water is like all else based upon the doctrine of The Conservation of Energy.

To be specific, Bernoulli's Law of 1738 states that along any particular hydrodynamic streamline the sum of gravitational potential energy, kinetic energy and compressive pressure energy is the same, though these energies can and do transact internally.

Bernoulli's Law may be written[12]:-

$$E = \frac{v^2}{2} + \Psi + \frac{p}{\rho} = K_1$$

**Equation 29**

where v is Stream Velocity; $\Psi$ is Gravitational Potential; p is Pressure and $\rho$ Fluid Density; and $K_1$ is a Constant specific to the streamline.

This statement is for incompressible fluid with no operational viscosity ( i.e. uninfluenced by boundary drag ). Therefore, pressure p should be viewed as dynamically coupled to velocity in Venturi effects.

Re-expression of Equation Twenty-Nine gives:-

$$\frac{1}{2}\rho v^2 + \rho g z + p = K_2$$

**Equation 30**

where Elevation z is positive above some datum and is much less than the planetary radius, which latter is assumed constant over the length of the streamline. In other words, g is assumed constant.

For an open channel flow, further simplification for the interconvertibility of potential and kinetic energy is feasible:-

$$\rho\left(\frac{v^2}{2} + gz\right) = K_3$$

**Equation 31**

and Equation Thirty-One lends a convenient basis for the description of roman wheels, Pelton wheels and other open-nappe hydraulic machines that work under atmospheric pressure.

The Bernoulli Equation requires emendation to reflect the geometrical and dynamical realities of a water-turned wheel.

Jean-Charles, Chevalier de Borda[13], applied Bernoulli energy-conservation principles to the roman wheel and in 1767 showed that because the overshot wheel exploited both the kinetic and potential energy of a given water parcel, whereas the undershot wheel used only the relatively weak kinetic effect, the overshot was about twice as efficient as the undershot.

In particular, for an overshot wheel:-

$$E_f = K\left(H_g - h_w\right) + \frac{Kh}{2}$$

**Equation 32**

where $E_f$ is the Maximum Effect ( measured as head in $L^1$ ); $H_g$ is the Distance Water Falls on the Wheel; $H_w$ is the Hydraulic-Height Determined Wheel Velocity Head; h is the Water Velocity Head; and K is a constant.

In a more practical idiom, Borda went forward to specify the overshot wheel's Energy Output, E, as :-

$$E = \gamma v = mgH_g + \frac{1}{2}mV^2 - \frac{1}{2}m(V - v)^2 - \frac{1}{2}mv^2$$

**Equation 33**

$\gamma$ is the Specific Weight of Water; v the Water Velocity within the Wheel Periphery; V is the Impact Velocity of the water as it hits the wheel; $H_g$ the Gravity Head is as aforementioned the effective Height of Fall in the fluid acting upon the wheel.

Analysis of the RHS of Equation Thirty-Three readily explains that output energy is the sum of potential and kinetic input energies with two losses: The third term of the RHS represents the loss of energy due to the impact of water against the wheel and the fourth term the loss due to the velocity remaining to the water on leaving the wheel ( i.e. an effective *upstream* flow of water at wheel-periphery velocity, v ).

Equation Thirty-Three can be simplified to head surrogates of energies as:-

$$E = K(H_g + h - h_i - h_w)$$

**Equation 34**

where $h_i$ is the Impact Head and $h_w$ the Wheel Velocity Head.

Except for the late turbine at Stone Mill, all of the Moddeshall Valley wheels were generically overshot and several were pitchbacks, a contrarotating variant designed to mitigate losses by eliminating Borda's fourth term of upstream-travelling exhaust water. By sending the spent water in a downstream direction, pitchbacks also optimised the performance of wheels when their tailraces were flooded.

The Denny Analysis

Denny[14] analysed an ideal overshot wheel in terms of Shaft Torque, $\tau$.

He described an ideal wheel with pivoted buckets of equilateral triangular section, each of which received a parcel of water of mass $\Delta m$ when it reached Top Center. The buckets processed spill-free until reaching Rotational Angle $\theta_1$, whereupon they were toppled perfectly to discharge their load to the tailrace.

Denny defined that:-

$$\Delta m = \frac{\rho Q}{\omega}\Delta\theta$$

**Equation 35**

where $\rho$ is Water Density; Q is Flow Rate and $\omega$ is the Angular Velocity of the Wheel ( in Radians/second ). $\Delta\theta$ is an Incremental Change in Bucket Angular Position. Furthermore, $\Delta t$, the Time to Fill the Bucket when multiplied by $\omega$ provides $\Delta\theta$.

The Number of Buckets mounted, n, was formalised by:-

$$n = \frac{2\pi}{\Delta\theta}$$

**Equation 36**

Denny derived the following expression for Shaft Torque, $\tau$:-

$$\tau = \frac{\rho g Q R}{\omega} \int_0^{\theta_1} Sin\theta d\theta = \frac{\rho g Q R}{\omega}\left[1 - Cos\theta_1\right]$$

**Equation 37**

R is the Wheel Radius that should be taken to be the shortest distance between the center of waterwheel rotation and the centers of mass of the ( identical ) buckets.

Accordingly, at steady state:-

$$\varpi = \frac{\gamma Q R}{\tau}\left[1 - Cos\theta_1\right]$$

**Equation 38**

Under such a condition the Impact Energy *per bucket* is:-

$$\Delta E_{in} = 2gR.\Delta m$$

**Equation 39**

and the Input Power *per bucket* is:-

$$\Delta P_{in} = \Delta E_{in}.\frac{\varpi}{\theta_1}$$

**Equation 40**

Defining Output Power as the product of steady angular velocity and $\tau_L$, the Load Torque, Denny moved forward to define the Efficiency, $\varepsilon$, of a given wheel as:-

$$\varepsilon \equiv Sin^2\left(\frac{1}{2}\theta_1\right)$$

**Equation 41**

Equation Forty-One seriously overestimates the efficiency of realistic bucket configurations.

Real roman wheels do not of course have pivoted buckets but fixed vanes and in overshot wheel variants these vanes are canted between the shrouds to form more or less watertight buckets.

These real buckets are subject to progressive spilling after some critical angle is reached, and also to significant centrifugal water spin-off if the angular velocity is too great.

Denny defined a general Loss Factor x(θ) to define the fraction of Δm lost at various θ positions. X(ω) and $X_1$(ω) are dimensionless Special Loss Factors tantamount to hydraulic efficiency metrics for a particular design and speed set-up. Denny extended his analysis to discriminate the torques due to potential and kinetic energies in these terms:-

$$\tau_g = \frac{\gamma QR}{\omega} \int_0^{\theta_{max}} Sin\theta d\theta = \frac{\gamma QR}{\omega} X_1(\omega)$$

**Equation 42**

and:-

$$\tau_w = \rho QR(v - \omega R)$$

**Equation 43**

$\tau_g$ and $\tau_w$ are respectively the Torques Due to Potential and to Kinetic Water Energy.

$\theta_{max}$ is the Centrifugal Critical Angle, measured with respect to zero being top center. v is Tangential Water Speed and ω is Peripheral Vane Speed.

To determine a value for $\theta_{max}$ apply the following equation:-

$$\theta_{max} = \cos^{-1}\left[\frac{R\omega^2}{g}\right]$$

**Equation 44**

In a direct parallel of Borda's Equation, Denny specified the overshot waterwheel Equation of Motion as:-

$$I\omega' = \tau_g + \tau_w - \tau_L - \tau_k$$

**Equation 45**

where, however, $\tau_L$ and $\tau_k$ are respectively Load and Friction Torques. I is The Moment of Inertia of the rotating system and ω´ is dω/dt.

Denny was able to show that taking both potential and kinetic inputs into account:-

$$\varepsilon = \frac{1 + Sin(\varphi)}{2 + \dfrac{v^2}{2gR}}$$

**Equation 46**

where φ is the Float Board Inclination To The Radius. Equation Forty-Six tends seriously to underestimate the efficiency of realistic bucket configurations.

Gate Power and the Denny Equations

$$P = \tau\omega$$
**Equation 47**

where P is the Power realised by the overshot waterwheel before losses, including those due to $\tau_L$ and $\tau_k$.

If we assume that $G \equiv P$ we may write:-

$$G = \omega\left(2\tau_g + \tau_w\right)$$
**Equation 48**

Substituting Equations Forty-Two and Forty-Three into the RHS we obtain:-

$$G = \omega\left(2\frac{\gamma QR}{\omega}X_1(\omega) + \rho QR(v - \omega R)\right)$$
**Equation 49**

Therefore:-

$$G = 2\rho g QR.X_1(\omega) + \rho QRv\omega - \rho QR^2\omega^2$$
$$= \rho QR\left(2g X_1(\omega) + v\omega - R\omega^2\right)$$
**Equation 50**

Because the bracketed terms constitute a quadratic equation there is an optimal rotational speed at which output power is maximised for any given Discharge and Wheel Radius.

So the best speed of an overshot wheel is not "as slow as practicable" as is sometimes counselled in literature.

In terms of the modelling spreadsheets applied, Equation Fifty was re-phrased as:-

$$G = P_g + P_w = 2\omega\tau_g + \omega\tau_w = \omega\left(2\tau_g + \tau_w\right)$$
$$= \omega\left[2\left(\frac{\gamma QR_E}{\omega}\right).REF + \rho QR_E\left(v - \omega R_E\right)\right]$$
$$= 2\gamma QR_E.REF + \omega\rho QR_E\left(v - \omega R_E\right)$$
$$= \rho QR_E\left[2g.REF + \omega\left(v - \omega R_E\right)\right]$$
**Equation 51**

where the Effective Wheel Radius, $R_E$, is given by:-

$$R_E = \frac{R + r}{2}$$
**Equation 52**

## Some Traditional Power Metrics

Overshot or breast wheel power equations need to comprise two additive terms: One to describe the contribution of Water Gravitation in the descending buckets; the other to define the contribution of the water's kinetic energy as the nappe hits the TC bucket and the water is slowed by the obstruction presented.

In a well-managed situation the kinetic component of generated energy may be anything from a fraction of a percent to about thirty-five percent but is typically around the ten or fifteen percent mark.

To be precise:-

$$G = Gravitational\ Power + Impulse\ Power$$
**Equation 53**

A robust, general and useful engineering equation requires basic mathematical integrity. Specifically it requires dimensional identity across the equality and the dimensional homogeneity of its additive terms. It should also have a pedigree of valid geometrical and physical derivation, and assemble constants and variables that are reliably ascertainable.

We will focus only upon overshot waterwheel power equations that meet these basic criteria of credibility.

### Borda's Power Equation

The first tenable power relation for the output of an overshot waterwheel is due to Jean Charles, Chevalier de Borda[13].

It is:-

$$P = K(H - h) + \frac{1}{2}Kh$$
**Equation 54**

An earlier equation, identical except for the coefficient of the kinetic term, was presented by Leonard Euler. This, however, was incorrect because the theoretician had failed to appreciate that the spouting nappe of water nearly halves its cross-sectional area as it leaps through space before hitting the bucket: A phenomenon called *vena contracta*.

In the context of Borda's Equation P is the Developed Power, H is Gravitational Head ( considered as useful fall in the form of wheel diameter ) and h is the Impulse Head or Kinetic Head due to Water Entry Velocity.

K is Weight of Water passing per Second ( force/time ) and is computed as:-

$$K = \frac{\rho g V_f}{t_f} = \frac{\gamma V_f}{t_f}$$
**Equation 55**

Where $\gamma$ is the Specific Weight of Water, $V_f$ is the Bucket Fill Volume $L^3$ ) and $t_f$ is the Bucket Presentation Duration ( T ).

Borda's Equation is a high-quality analytical construct and yields computational power estimates virtually identical to those of the present Denny[14]-Warren Model.

For comparative purposes, Borda's Equation was incorporated into the Denny-Warren Model is these terms:-

$$P_B = \frac{\gamma V_f}{t_f} \times 2R_E.REF + \frac{1}{2}.\frac{\gamma V_f}{t_f}.\frac{v^2}{2g}$$

$$= \frac{\gamma V_f}{t_f}\left(2.REF + \frac{v^2}{4g}\right)$$

**Equation 56**

The Franklin Institute Equation

The Franklin Institute[15] Power Equation is a theoretico-empirical equation following the experimental spirit of the eighteenth-century English civil engineer John Smeaton who catalogued waterwheel efficiency tests in a monograph.

The Franklin Institute Formula is:-

$$P = 0.90K(H - h) + 0.28Kh$$

**Equation 57**

H is the Total Head, (H-h) is the Gravity Fall and h the Impact Head ( i.e. $v^2/2g$ ). K is the Weight of Water per Second as defined above.

With these appropriate substitutions, the Franklin Institute Equation was computed as:-

$$P_{FI} = 0.9K(2R_E.REF) + 0.28K\frac{v^2}{2g}$$

**Equation 58**

A further reduced form is of course available as:-

$$P \approx 0.28K\left(6.428571.R_E.REF + \frac{v^2}{2g}\right)$$

**Equation 59**

# PART IV
## POWER RATINGS OF INSTALLED AND HISTORICAL
## MODDESHALL VALLEY HYDROPOWER WHEELS

To assess the power of the Moddeshall System mills I made several assumptions:-

(a)    That all nine of the wheels that survived into industrial times were plank wheels whose buckets had infinitely thin floats and shrouds, and no rising boards.

(b)    That the radius of gyration was constant around the circumference, and that it was the mean of inner and outer radii.

(c)    That water nappes could be made to leap into wheels at up to three meters per second.

(d)    That all wheels were, or could be replaced by, overshot wheels.

These assumptions are known to be unrealistic or just plain wrong. I consider, however, that these errors and uncertainties are unimportant from the point of view of practical water engineering, even in cumulate.

In addition, certain assumptions were made regarding environmental factors. These include, but are not limited too:-

(e)    That wheel powers can be associated with head reach stream powers, despite spacial disjunctions: Stream powers are associated with specific five-meter contours; wheel heads are at arbitrary elevations.

(f)    That contemporary mean discharge operated the wheels and that it is the product of areally-summated rainfall averages with their catchment subareas.

Despite these several vagaries and the myriad others we have not addressed, or even apprehended, it is possible to elicit some useful principles of the history and potential of Moddeshall Valley hydropower.

We shall not here dwell upon the mathematics or physics of wheel operation. The physics has been outlined in the previous section, and the mathematics can be found in Technical Appendices A and B.

Rather we shall focus upon tenable modelling of wheel power production and the power relations of the Moddeshall System streams and their mills.

## Known Facts about Moddeshall Mill Wheels

My knowledge of the type, dimensions and employments of Moddeshall wheels is taken from the Barry Job monograph "Watermills of the Moddeshall Valley"[1], privately published by Dr Job.

His information, and my inferences from his diagrams and photographs, are summarised in Data Appendix C.

## Bridging gaps in My Knowledge

In the case of four of the wheels I was able to assess the relative bucket depth ( as shroud annulus depth ) by reference to photographs. Knowledge of the absolute wheel diameters then allowed actual bucket depths accurately to be estimated.

The four wheels concerned were the existing devices at Splashy, Mosty Lee and Coppice Mills, and another at Ivy Mill shown in an extant photograph.

The bucket designs of the other five wheels is unknown to me, so I sought to make reasonable assumptions about their design, believing that if anything had influenced millwrights of the past, other than workshop economy or convenience, then that thing would have been the Flow Rate of the lade in question.

Figure Eight plots wheel Radius/Annulus Ratio against lade Discharge ( assumed river Flow Rate ) for the four known wheels.

**Discharge versus Radius/Annulus Ratio**

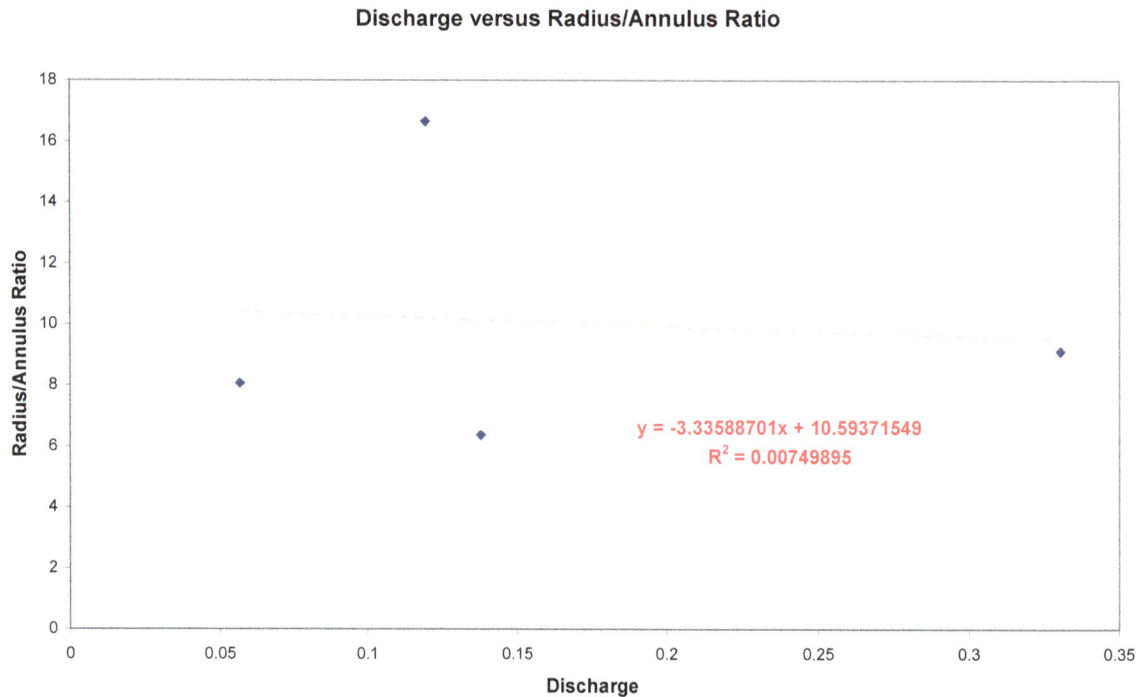

$$y = -3.33588701x + 10.59371549$$
$$R^2 = 0.00749895$$

**Figure Eight**
**Bucket Depth No Function of Flow**

You can see that there is no significant relationship.

Accordingly, I adopted the mean Radius/Annulus Ratio of 10.05630760 for all five wheels and computed historic bucket depths on that basis.

A second crucial model parameter was unknown in the five cases.

This is the Number of Buckets per Wheel. Again, for the four photographed wheels this was readily apparent by counting the buckets in a $\pi/2$ quadrant and simply multiplying by four.

In my experience, it is rare for the Bucket Count to be other than $2^m$ or $2^{m-1} \times 3$, where m is an arbitrary integer. I assumed that Moddeshall wheels would have such an even count, knowing the existing Splashy, Mosty Lee, Ivy and Coppice wheels respectively to have 24, 56, 84 and 108 buckets.

Figure Nine plots Bucket Count against Discharge for the four seen wheels and you can compute that the count is about 64% determined by design flow rate.

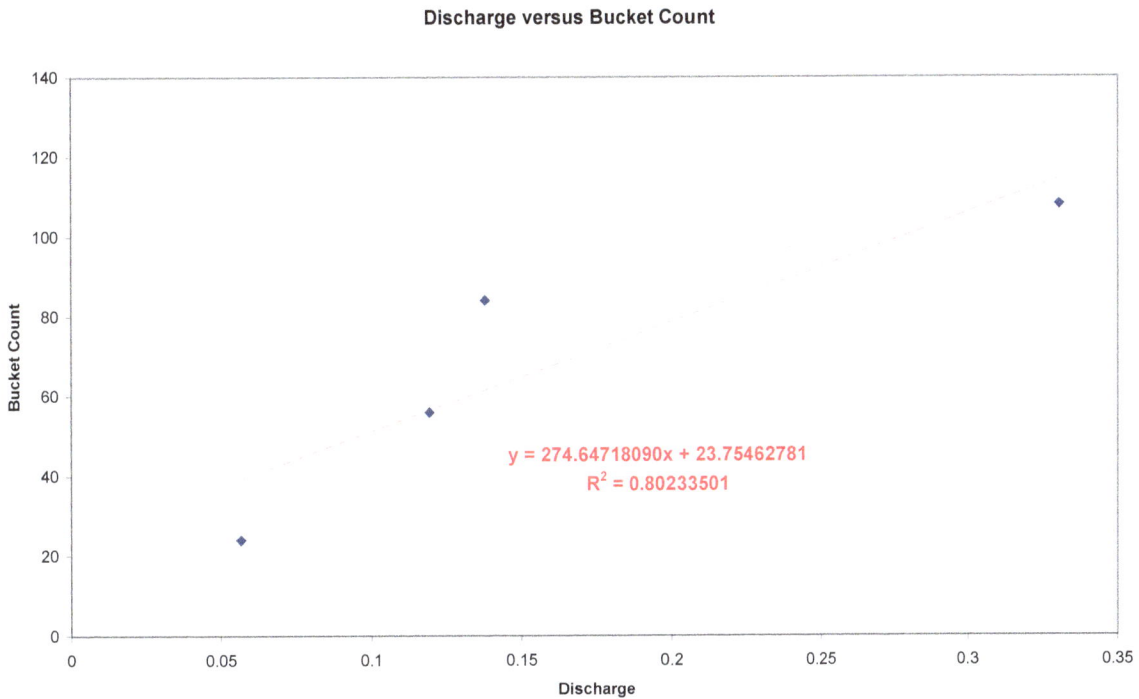

Discharge versus Bucket Count

$$y = 274.64718090x + 23.75462781$$
$$R^2 = 0.80233501$$

**Figure Nine**
**Installed Bucket Count Increases with Discharge Considered**

In consequence of this, I used the fitted linear polynomial to compute a notional Bucket Count for the five undocumented wheels and adjusted it to the nearest $2^m$ or $2^{m-1} \times 3$ value, consonant with a sensible aperture size: About 17 centimeters in the narrowest instance.

Tables Nine and Ten summarise the relevant wheel design statistics, respectively in Imperial and Metric units.

| Mill Name | Grid Reference | Head Elevation (meters) | Dimensions (Feet) | | | | Number of Buckets | Radius/Annulus |
|---|---|---|---|---|---|---|---|---|
| | | | Diameter | Outer Radius R | Width | Bucket Annulus (feet) | | |
| Boar | SJ926368 | 169 | 20 | 10 | 4.5 | | | |
| Splashy | SJ919366 | 143 | 16 | 8 | 5.5 | 0.99 | 24 | 8.06 |
| Ochre | SJ918365 | 137 | 12 | 6 | 9 | | | |
| Mosty Lee | SJ918362 | 136 | 18 | 9 | 6.5 | 0.54 | 56 | 16.67 |
| Wetmore | SJ916360 | 140 | 21 | 10.5 | 7 | | | |
| Ivy | SJ916355 | 129 | 19 | 9.5 | 6 | 1.49 | 84 | 6.37 |
| Hayes | SJ912351 | 113 | 20 | 10 | 6.25 | | | |
| Coppice | SJ908347 | 116 | 20 | 10 | 6.5 | 1.10 | 108 | 9.13 |
| Stone | SJ905340 | 98 | 24 | 12 | 5.75 | | 68 | 10.06 |

**Table Nine**
**Wheel Design Statistics in Imperial Units**

# Table Ten
# Wheel Design Statistics in Metric Units

| Mill Name | Grid Reference | Dimensions (Meters) | | | | Number of Buckets n | Bucket Annulus (meters) | Radius/ Annulus | Cumulative Discharge (m³s⁻¹) | Nominal Stream Power at Intake (watts) | Raw Bucket Count | Adjusted Bucket Count |
|---|---|---|---|---|---|---|---|---|---|---|---|---|
| | | Diameter | Outer Radius R | Inner Radius r | Width W | | | | | | | |
| Boar | SJ926368 | 6.10 | 3.05 | 2.74 | 1.37 | 24 | | | 0.04928844 | 19525 | 37 | 24 |
| Splashy | SJ919366 | 4.88 | 2.44 | 2.14 | 1.68 | 24 | 0.30 | 8.06 | 0.05665626 | 14946 | 39 | |
| Ochre | SJ918365 | 3.66 | 1.83 | 1.65 | 2.74 | 48 | | | 0.09918652 | 14946 | 51 | 48 |
| Mosty Lee | SJ918362 | 5.49 | 2.74 | 2.58 | 1.98 | 56 | 0.16 | 16.67 | 0.11940882 | 14946 | 57 | |
| Wetmore | SJ916360 | 6.40 | 3.20 | 2.88 | 2.13 | 60 | | | 0.1304819 | 14946 | 60 | 60 |
| Ivy | SJ916355 | 5.79 | 2.90 | 2.44 | 1.83 | 84 | 0.45 | 6.37 | 0.1378789 | 11068 | 62 | |
| Hayes | SJ912351 | 6.10 | 3.05 | 2.74 | 1.91 | 96 | | | 0.14373416 | 5409 | 63 | 96 |
| Coppice | SJ908347 | 6.10 | 3.05 | 2.71 | 1.98 | 108 | 0.33 | 9.13 | 0.33045183 | 12143 | 115 | |
| Stone | SJ905340 | 7.32 | 3.66 | 3.29 | 1.75 | 128 | | | 0.34153125 | 26244 | 118 | 128 |
| Means | | | | | | 70 | | 10.06 | | | | |

0 Degree Coeff    23.755
1 Degree Coeff    274.650

RED VALUES are estimates

The rubricated values are my interpolations.

A third unknown factor is the Gate Aperture. In all but the last two mills, the Coppice and Stone Mills, the Gate ( Sluice ) Aperture was assumed to be $0.06m^2$. These values were chosen to approximate one to three meters per second of flow: Rather brisk for the Moddeshall context, but possibly of little import for power genesis.

Loading Policy

Modelled wheels loads were minimised in terms of manual heuristic fill depths, consistent with the evasion of mathematical "skipahead" anomalies.

In practice this implies that kinetic energy inputs were minimal whilst gravitational output was fully developed: At least in terms of my computer spreadsheet modelling based upon the Denny-Warren equation system.

The Wheel Performance Results for the nine historically-documented Moddeshall wheels are presented in Data Appendix D.

Figure Ten illustrates the model Fill Depths, $d_{min}$, and also the approximate Half Depths for all nine mills. It can be seen that optimal $d_{min}$ is about 12% to 17% of bucket depth in most cases: Extremes are about 5% for Ivy Mill and 17% for Mosty Lee, though human error probably accounts for much of the variation.

Minimum Depth of Fill and Bucket Half Fill

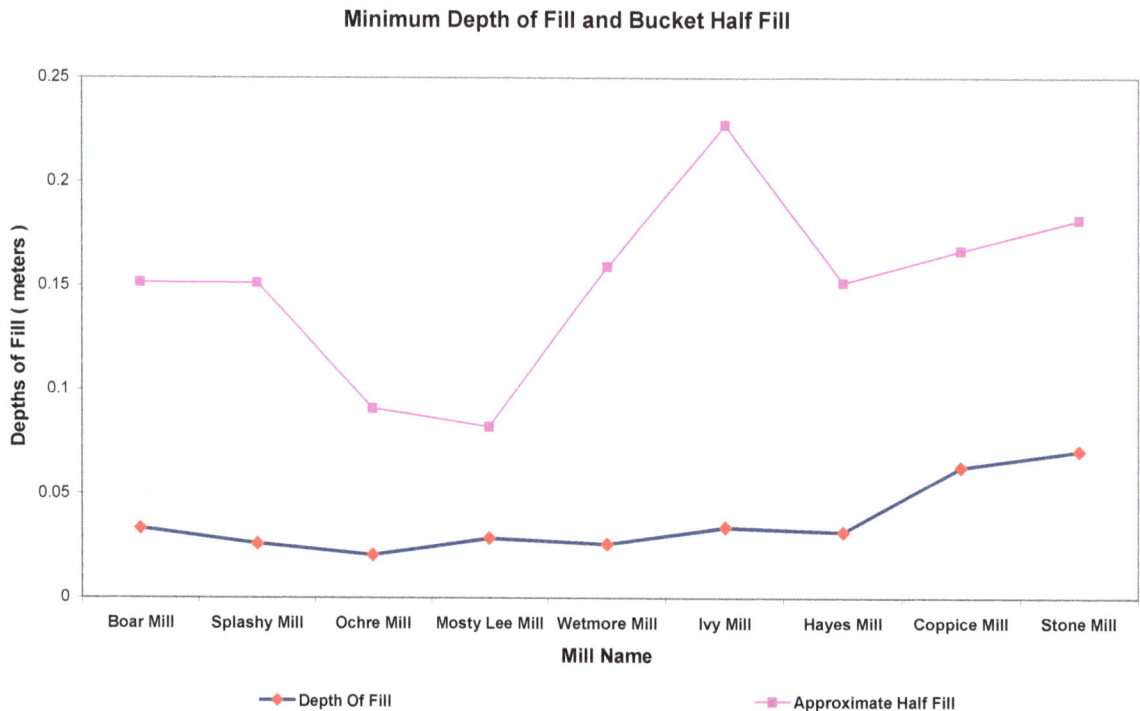

**Figure Ten**
**Depth of Fill for the Moddeshall Mills**

## Cycle Time

There is strong analytic evidence that there is a finite optimal overshot wheel Angular Speed, $\omega$. It is of course possible to equivalate $\omega$ as Angular Velocity in Hertz or indeed as Cycle Time in seconds; reciprocal angular velocity.

Millers' craftlore says that wheel speed should be "as slow as practicable" and this doctrine is propagated in a lot of learned literature.

Figure Eleven depicts the Cycle Time. It varies from as slow 22 or 23 seconds in the headwaters of Boar Mill to about 6.5 seconds at Ochre Mill.

Remarkably, the average Cycle Time varies only in the range 6.5 to 9 seconds throughout the long range of The Scotch Brook dampening to about 7.5 seconds for the final mills.

This behaviour furnishes strong empirical evidence for a finite optimal wheel speed.

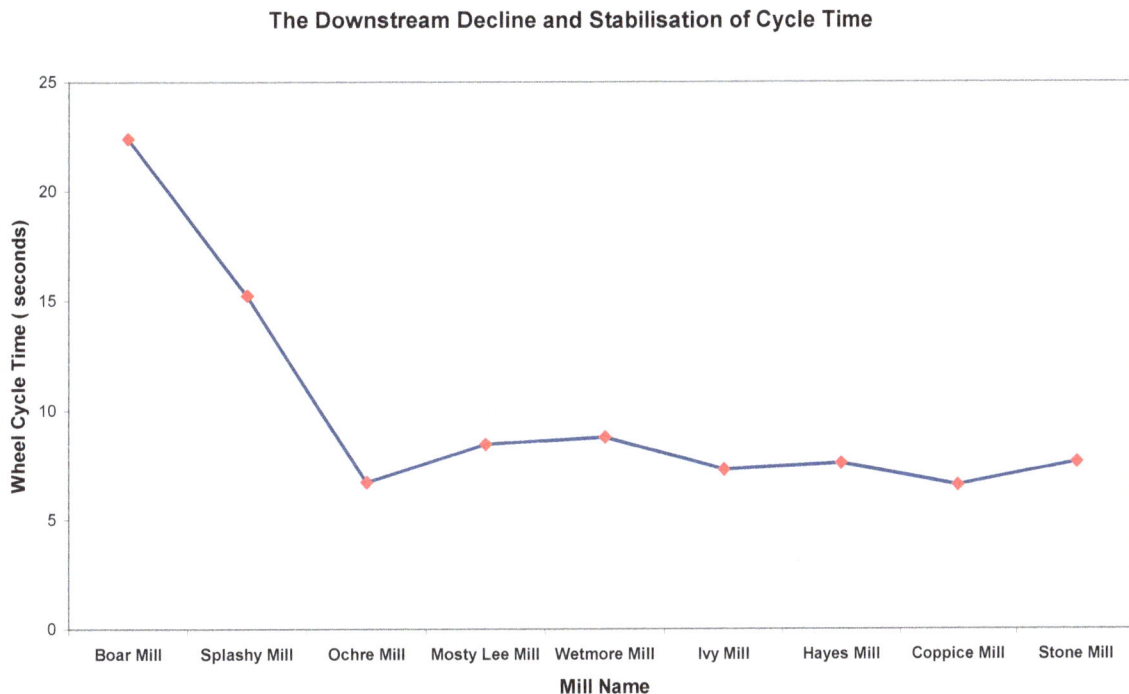

**The Downstream Decline and Stabilisation of Cycle Time**

**Figure Eleven**
**Modelled Moddeshall Wheel Cycle Times**

## The Retentional Efficiency Fraction

The Retentional Efficiency Fraction, REF, encapsulates the history of bucket water load retention as it processes from the inlet point at or near wheel Top Center until and beyond the point of complete emptying at Bottom Center.

REF is essentially the same as the hydraulic efficiency of an overshot waterwheel, without of course taking mechanical or electrical energy losses into account.

REF shows a steady and consistent linear decline as we progress downstream, from a high of 89% at Boar Mill to only 62% at Stone Mill, where the Fourneyron turbine installed circa 1860 must have been a much superior choice of engine.

This is strong evidence that, for overshot waterwheels, efficiency is inversely related to discharge, so that they are a good choice for little streams, and an ill choice for big rivers.

The downstream decline of REF is illustrated by Figure Twelve.

The Downstream Decline of REF

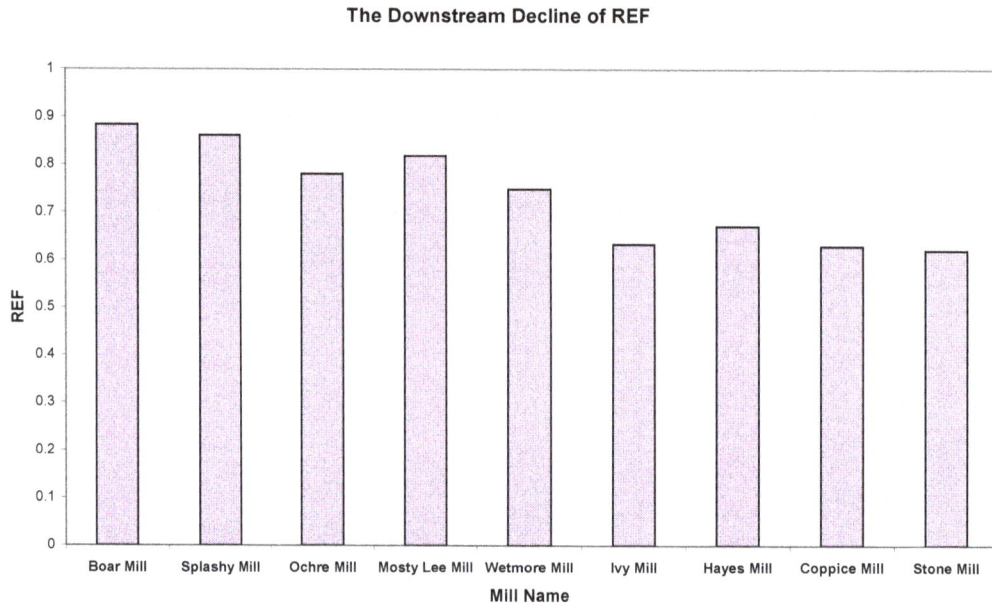

**Figure Twelve**
**The Downstream Decline of REF**

## Power Profiles

The doctrine of Comparative Power Profiles is based upon The Conservation of Energy and the idea that a waterwheel cannot transduce more energy than is engendered in the stream between the wheel and whatever last removed the energy the stream contained.

In the last century and previously, dam constructors erred to their cost and to the more final calamity of sublaqueous populations.

We need to share power with a river in a spirit of courteous respect. Otherwise the stream shall sulk by shedding its sediment load upon our installations; express its anger in avulsion; or with sullen contempt sweep away our feeble works.

Clearly, it is not in our long-term interest to abuse the watercourse by over-exploitation, or by inappropriate or wasteful minor uses.

Analysis of the Moddeshall System Comparative Profiles is partly obscured by mill crowding and in particular by the fact that four mills sit upon a single power reach between the 135 and 160 meter OD contours. The total yield of this reach is near to 14946 watts and the four mills must have shared this amongst them, though the head mill ( Splashy ) could "borrow" energy from its superior reach, if any remained there.

Figure Thirteen illustrates the situation.

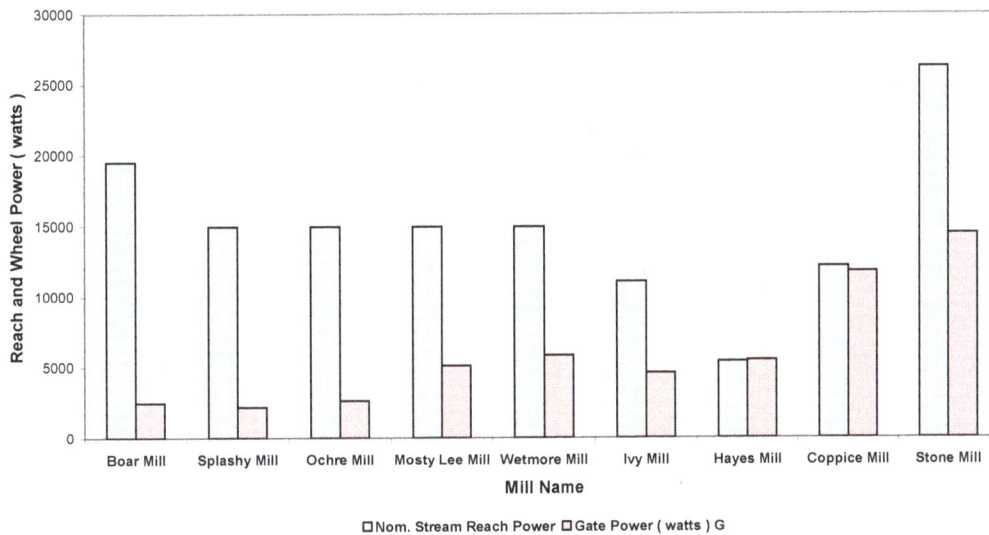

Comparative Power Profiles
Of Available Stream Reach Power and Installed Wheel Power

**Figure Thirteen**
**Reach and Wheel Comparative Power Profiles**

The blue-patterned columns represent the power engendered throughout a mill head reach by virtue of flow augmentation and water gravitation, epitomised by the concept of "hydraulic head".

The red-patterned columns are the subject wheel powers.

In a well-managed watercourse the wheel powers are always considerably less than the natural reach powers.

You can see that the old Boar Mill location had ten times more natural power than men consumed, whilst the next four mills, of steadily increasing wheel power, sapped the same central reach, though Splashy Mill was subsidised by the power unused by Boar Mill.

We know from history that there was a complex system of power sharing at the confluence of the Moddeshall and Scotch Brooks that definitely involved co-ordination between Splashy and Ochre Mills. We further infer that Mosty Lee and Wetmore were probably involved, as most likely were retention dams in the headwaters of Scotch Brook and an independent millpond above Ochre Mill.

Further down Scotch Brook, Ivy Mill took less than half the power of its head reach, and subsidised Hayes Mill, which used more power than its own head reach could possibly have supplied.

Coppice Mill used nearly as much power as arose in its headwaters.

Like Boar at the top of the system, Stone Mill at the bottom is a great anomaly. Stone Mill consumed only fourteen of the twenty-three kilowatts available to it, though it is known to have had a second wheel for nocturnal use which may have been run in tandem with the first wheel under propitious flow conditions. In particular, an upstream millpond was probably allowed to recuperate during the day to provide an augmented flow rate for night work. Also, the late Fourneyron turbine may have raised more power.

Figure Fourteen helps to clarify the situation of the Central Mills ( Splashy, Ochre, Mosty Lee and Wetmore ) by consolidating their demands beside the available fifteen or so kilowatts of the 135-160 vertical meters reach.

**Grouped Comparative Power Profiles**
**Of Available Stream Reach Power and Installed Wheel Power**
*Grouped Mills are Splashy, Ochre, Wetmore and Mosty Lee*

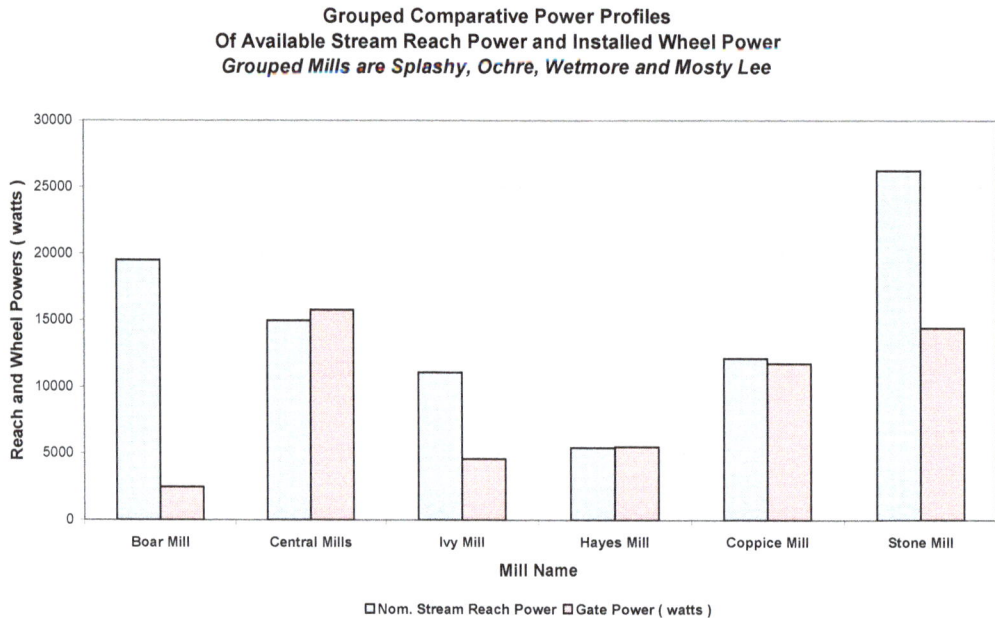

□ Nom. Stream Reach Power □ Gate Power ( watts )

**Figure Fourteen**
**Reach and Grouped Wheel Comparative Power Profiles**

It is now clear that the Central Mills are marginally overbalanced and partially depended upon the vigor of water entering from the higher altitudes.

Overall, the installed mills abstracted 61% of the natural power of the Moddeshall Stream System, with only the extreme mills at Moddeshall village and Stone having potential to spare.

Table Fifteen summarises the Reach and Wheel Powers for the nine industrial Moddeshall Valley mills.

| Name | | Boar Mill | Central Mills | Ivy Mill | Hayes Mill | Coppice Mill | Stone Mill | Totals |
|---|---|---|---|---|---|---|---|---|
| Nom. Stream Reach Power | | 19525.20022 | 14946.05417 | 11068.20193 | 5409.338021 | 12142.74799 | 26243.92343 | 89335.46576 |
| Gate Power ( watts ) | G | 2474.114048 | 15765.69387 | 4570.36907 | 5477.19189 | 11760.9488 | 14444.70718 | 54493.02486 |
| | | | | | | | | |
| Explotation (%) | | | | | | | | 60.99819864 |

**Table Fifteen**
**The Moddeshall System Mills Grouped Power**

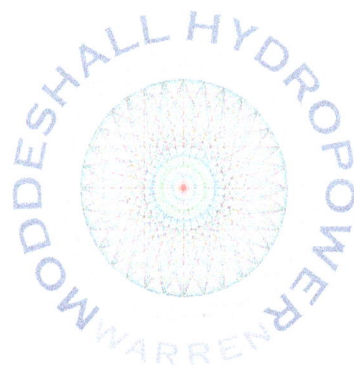

# PART V
# CONCLUSIONS

Some have remarked that there is an irrational prejudice against the roman wheel, many regarding it as the embodiment of inefficiency and archaism.

For sure the word *roue* has several evil echoes, especially for Frenchmen, and none would have been lost on Euler, Borda and their correspondents. And yet in different lore and recollection the waterwheel is a perpetual token of the cycles of life and rather than a remembrance of excruciating death becomes a symbol of Resurrection, like the Cross that lends it form.

Overshot waterwheels, and the other gravity wheel variants of that type, are, when intelligently designed and prudently piloted, highly efficient and reliable sources of power. In such regards they are, in their place, as good as any modern turbine.

In suburban or wooded catchments, and the Moddeshall Valley is both suburban and wooded, overshot wheels are especially commended by their structural robustness. Floating logs, bottles, gas cylinders and other debris not arrested by the headwork tends to vault harmlessly over the arc and fall into the tailrace without expensive damage or interruption of service. Costly, labor-intensive and power-draining trash gratings are seldom needful, and few thieves and vandals are stupid enough to approach a turning waterwheel, though malefactors have rendered passive Stoke-on-Trent water facilities unusable.

Where the roman wheel defers to the water turbine is in the modern need for a high-speed, low-torque revolution to optimise electrical equipment. But with twenty-first century torque converters, static inverters and electronic control even this is not the impediment it was. At the Falls of Aberdulais in the lovely Neath Valley a modern steel suspension simple overshot wheel hub-drives a conventional alternator to yield a steady twenty-five kilowatts, enough to satisfy the power requirements of the local National Trust visitor center. Beneath this picturesque wheel, a hidden Kaplan turbine raises a further 200Kw for sale to the National Grid.

For the Moddeshall Valley, our computer spreadsheet model showed that the nine documented mills almost exactly matched to local power engendered by the stream, the exceptions being the Boar and Stone Mills at the extremities of the system. These two utilised only half of the available local power, though there is reason to think that nineteenth-century owners understood this and took steps to lay on turbines or intermittent wheels.

The fact that the Moddeshall Mills took only 61% of the available waterpower probably meant that enough celerity remained to prevent the stream becoming a stagnant ditch, which is pretty well what it now is, at least at Stone.

It would be good if the Moddeshall wheels were modernised and reinstated to give maybe forty or fifty kilowatts to the riparian community. Local and national engineering and industry would modestly be encouraged and permanent employment could be provided for a bailiff and a mechanic, with occasional work for electricians.

Table Sixteen presents the expected revenues of the Moddeshall System wheels were they re-employed to generate electric power. The Source Value row shows the annual monetary yield at each documented mill site in British Pounds Sterling ( November 2009 ). The Generation and Export Tariff figures assume the 2010 implementation of the UK Government's "Clean Energy Cashback" policy structure at its proposed levels. This policy subsidises microhydro implementations to the tune of around seventy percent.

"Clean Energy Cashback" brings in about £93,605 per annum for a re-installed system, whilst for comparison the current commercial retail value of the electricity is around £54,934 assuming a charge of 11.5 pence per kilowatt-hour to domestic premises.

At a *pro rata* yield the 3033.53 km$^2$ of the historical County of Staffordshire would supply about 16.71MW of stream power worth about £7.325million at 2009 retail values, whilst the entire UK would produce about 1.33GW worth £1.34milliard. Halve these figures realistically to estimate the electric power potential.

The roman wheel has an honorable place in history and will be rehabilitated. It has asked little and given much. It has spared the weakest the ceaseless grinding labor of food processing and made possible the standing army and the sustained campaign. Without it the Roman Empire would not have arisen, and doubt exists whether Chinese civilisation would have established either, for the people of China developed a very similar technology during the Iron Age.

Steve Dalton gives a useful checklist of the technologies and issues surrounding modern microhydro in his PowerPoint presentation "Low Head Hydro Technologies and the need for greater uptake and implementation on the UK"[16].

Some of the future research avenues that have proclaimed themselves include:-

(a)     The Determinants and Computation of Optimal Wheel Speeds.
(b)     The Implications for Power and Efficiency of Serial and Parallel
        Wheel Working, at both Local and Catchment Scales.
(c)     The Feasibility of Compound Systems.
(d)     Efficient Torque Conversion.
(e)     The Use of Rimdrive Systems to Yield 50Hz Electric Current
        with Minimal Loss, including using the Wheel Itself as a
        Generative Rotor.
(f)     Fish-friendly Installations and Working.
(g)     The Effects of Over-Exploitation on Drainage and Ecology.
(h)     Automatic Orchestration of Catchment Wheel Behaviour for
        Real-Time Meteorological and Hydrological Changes.

This is not of course an exhaustive list.

I started this essay thinking it would be the work of a few hours. So far, it has occupied three months, and there is still much to do, for I now understand that there is more to know of the simple waterwheel than a man can comprehend.

The Moddeshall Valley gave me great pleasure and made me think. For myself, I can ask no more of it. As I got back into my aged Volvo 460 sports sedan and reached fifty miles per hour through the early autumn sunlight I was conscious that I alone was generating seventy-four kilowatts.

| Name | Boar Mill | Splashy Mill | Ochre Mill | Mosty Lee Mill | Wetmore Mill | Ivy Mill | Hayes Mill | Coppice Mill | Stone Mill | Totals |
|---|---|---|---|---|---|---|---|---|---|---|
| Nom. Stream Reach Power ( watts ) | 19525 | 14946 | 14946 | 14946 | 14946 | 11068 | 5409 | 12143 | 26244 | 89335.4658 |
| Gate Power ( kW ) | 2.474 | 2.191 | 2.641 | 5.107 | 5.828 | 4.570 | 5.477 | 11.761 | 14.445 | 54.4930249 |
| Exploitation (%) | | | | | | | | | | 60.9981986 |
| Hours per Year | 8766 | 8766 | 8766 | 8766 | 8766 | 8766 | 8766 | 8766 | 8766 | |
| Electricity Yield (Kwh/annum ) | 21688 | 19202 | 23149 | 44764 | 51087 | 40064 | 48013 | 103096 | 126622 | 477686 |
| Generation Tariff (£/Kwh) | £0.17 | £0.17 | £0.17 | £0.17 | £0.17 | £0.17 | £0.17 | £0.12 | £0.12 | |
| Export Tariff (£/Kwh) | £0.05 | £0.05 | £0.05 | £0.05 | £0.05 | £0.05 | £0.05 | £0.05 | £0.05 | |
| Source Value (£/annum) | £4,771.38 | £4,224.49 | £5,092.75 | £9,848.03 | £11,239.19 | £8,814.05 | £10,562.87 | £17,526.40 | £21,525.79 | £93,604.95 |
| Retail Price (£/Kwh) | £0.115 | £0.115 | £0.115 | £0.115 | £0.115 | £0.115 | £0.115 | £0.115 | £0.115 | |
| Retail Value | £2,494.13 | £2,208.26 | £2,662.12 | £5,147.83 | £5,875.03 | £4,607.34 | £5,521.50 | £11,856.09 | £14,561.56 | £54,933.87 |

**Table Sixteen**
**Potential Electric Revenues in The Moddeshall Valley**

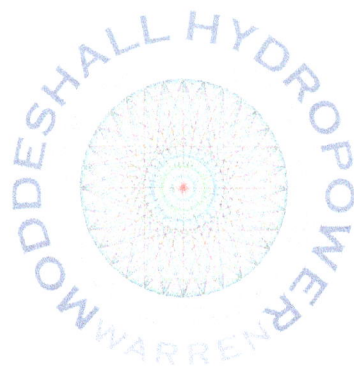

## PART VI
## REFERENCES

**Number**

0 **Standard Acceleration of Free Fall**
  **Water Density Equation**
  **Equatorial Radius**
  **Polar Compression**
  **Bernoulli Law**
  "CRC Handbook of Chemistry and Physics"
  1st Student Edition     1988
  Editor-in-Chief: Robert C Weast PhD
  CRC Press incorporated of Boca Raton FL
  ISBN 0-8493-0740-6


0 **Radius of an Ellipse**
  "Handbook of Mathematics"
  IN Bronshtein and KA Semendyayer
  Verlag Harri Deutsch of Thun     1979
  ISBN 3-87144-644-0
  p201
  973pp


0 **Water Temperature at Shrewsbury**
  "The prediction of river water temperatures"
  K Smith
  Bulletin des Sciences Hydrologiques
  Vol.26 Part1 March 1981


1 **Mills in the Moddershall Valley**

  Watermills of the Moddeshall Valley
  Barry Job
  Privately published by the Author
  Newcastle-under-Lyme     1995
  119pp


2 **BSI Stream Discharge Correction Table**

  USBR Water Measurement Manual: Third Edition: 2001
  United States Bureau of Reclamation
  Chapter 13: Special Measurement Methods in Open Channels
  Page 13-4
  http://www.usbr.gov/pmts/hydraulics_lab/pubs/wmm/wmm.html

### 3 Manning's Equation

Layman's Guidebook on How to Develop a Small Hydropower Site
European Small Hydropower Association     June 1998
266pp
Page 33
http://ec.europa.eu/energy/library/hydro/layman2.pdf

### 4 Multimap

http://www.multimap.com/maps/

### 5 Trent Catchment Isohyet Map
28006 - Trent at Great Haywood
Rainfall
http://www.nwl.ac.uk/ih/nrfa/spatialinfo/Rainfall/rainfall028006.html

### 6 EA Gauging Station Summary Sheets
The National River Flow Archive Station Summary Sheets
http://www.nerc-wallingford.ac.uk/ih/nrfa/station_summaries/sht.html

### 7,8 British Hydrometric Register
http://www.ceh.ac.uk/products/publications/documents/
HydrometricRegister_Final_WithCovers.pdf

### 9 Cresswell Groundwater Abstraction
STW_Drought_Plan_low_res_20060724121702.pdf
Section 5.4.7 Pages 87-89
http://www.stwater.co.uk/upload/pdf/STW_Drought_Plan_low_res_20060724121702.pdf

### 10 Water Density Equation
1st Student Edition     1988
Editor-in-Chief: Robert C Weast PhD
CRC Press incorporated of Boca Raton FL
ISBN 0-8493-0740-6

### 11 Catacomb Depiction of a Waterwheel
"Stronger than A Hundred Men"
Terry S Reynolds
The Johns Hopkins University Press
Baltimore and London
ISBN 978-0-8018-7248-8
pp453

12 **Bernoulli's Law**
"CRC Handbook of Chemistry and Physics"
1st Student Edition    1988
Editor-in-Chief: Robert C Weast PhD
CRC Press incorporated of Boca Raton FL
ISBN 0-8493-0740-6

13 **Borda Equations**
Jean Charles, Chevalier de Borda
"Mémoire sur les Roues Hydrauliques"
Académie des Sciences, Paris, Histoire
1767 ( pub 1770 )
Pages 278-80 cited on p239 of Reynolds

14 **Denny Analysis**
"The Efficiency of Overshot and Undershot Waterwheels"
Mark Denny
5114 Sandgate Road, Victoria, BC,  V9C 3Z2, Canada
2 December 2003
European Journal of Physics, V25 ( March 2004 ), Issue 2, pp193-202
The Institute of Physics Publishing

15 **Franklin Institute Power Equation**
"Report of Committee of the Franklin Institute of Pennsylvania, appointed May, 1829,
to Ascertain by Experiment the Value of Water as a Moving Power"
Franklin Institute Journals 11(1831);12(1831);13(1832);14(1832);31(1841) and 32(1841)
Various Passages
The Franklin Institute Of Pennsylvania, Philadelphia
cited on p255 of Reynolds

16 **Current Technologies Overview for Small Hydropower**
Low Head Hydro Technologies
and the need for greater uptake
and implementation in the UK
An Engineer's Perspective
by Steve E Dalton
Clean Energy Solutions Ltd
http://www.cleanenergysolutions.co.uk/presentation.pdf

17 Selling Microhydro Electricity to the UK Grid
"What is a feed-in tariff?"
Discussion of HMG "Clean Energy Cashback" Proposals for 2010
http://www.yougen.co.uk/selling-electricity-to-the-grid/

A        **Stream Power Equation**

Geomorphology
Richard J Chorley
Routledge      20 Dec 1984
ISBN 978-0416325904
648pp
Page 280


B        **Stream Power Equation ( with Valid Dimensions )**

Variability in discharge, stream power, and particle-size distributions
in ephemeral-stream channel systems
LJ Lane, MH Nichols, M Hernandez, C Manetsch, WR Osterkamp
International Association of Hydrological Sciences
Variability in Stream Erosion and Sediment Transport
Proceedings of the Canberra Symposium, December 1994
IAHS Publication Number 224, 1994
pp335-342


C        **British PE and AE Review for 2002**

http://www.nerc-wallingford.ac.uk/ih/nrfa/yb/yb2002/evaporation.html


D        **British Hydrometric Register**

http://www.ceh.ac.uk/products/publications/documents/
HydrometricRegister_Final_WithCovers.pdf


E        **The Limerinos Equation**

Guide for Selecting Manning's Roughness Coefficients
for Natural Channels and Flood Plains
United States Geological Survey Water-supply Paper 2339
Metric Version

http://www.fhwa.dot.gov/BRIDGE/wsp2339.pdf


F        "Stronger than A Hundred Men"
Terry S Reynolds
The Johns Hopkins University Press
Baltimore and London
ISBN 978-0-8018-7248-8
pp453

G        **Current Technologies Overview for Small Hydropower**
         "Low Head Hydro Technologies
         and the need for greater uptake
         and implementation in the UK"
         "An Engineer's Perspective"
         by Steve E Dalton
         Clean Energy Solutions Ltd
         http://www.cleanenergysolutions.co.uk/presentation.pdf

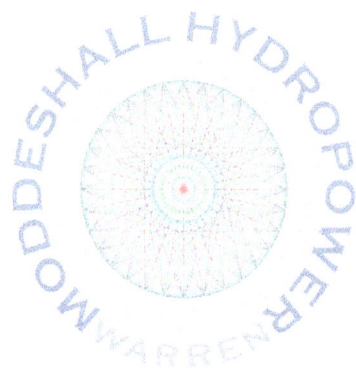

# TECHNICAL APPENDIX A
## ANALYSES OF BUCKET DESIGN AND FILLING GEOMETRIES

As Denny showed quite recently, the mass of water carried in the bucket of an overshot wheel is a critical factor in the wheel's power output.

The Mass, m, of water in a given bucket is the product of the Water Density, $\rho$, which is a regional constant, and the Volume, V, of water contained. Because a bucket is of uniform Cross-Sectional Area, A, it is enough for us to assess bucket capacity as a function of this wetted section and the bucket design section.

In formal terms:-

$$m = \rho V = \rho A W$$
**Equation A1**

where W is the Filled Bucket Width. The buckets of a roman wheel are of course all of the same wetted width, and in the final stages of industrial water wheel employment designers made waterwheels as wide as was structurally supportable in order to make the wetted profile as low as practicable and delay the bucket emptying as long as possible, since the power efficiency is a direct function of containment durations.

Later we will need to move forward to describe the changes in wetted section with Wheel Lapsed Angle, $\psi$, especially to identify the Spill Angle, $\psi_{spill}$, and the profile of water loss between the Spill Angle and the Bottom Center at $\pi/2$ radians, at which we may assume all the water to have been tipped out.

In the latter stages of industrial roman wheel design bucket profiles were forged in parabolic and other complex shapes in an attempt to optimise efficiency and economy but in the heyday of the overshot wheel, say the middle of the eighteenth century, wheels were built up of flat wooden planks.

Such plank wheels offer us a simplified analysis that throws light upon certain principles of wheel design and optimisation.

Figure A1 illustrates some useful parameters of plank wheel design and fill with regard to a bucked backed at Top Center:-

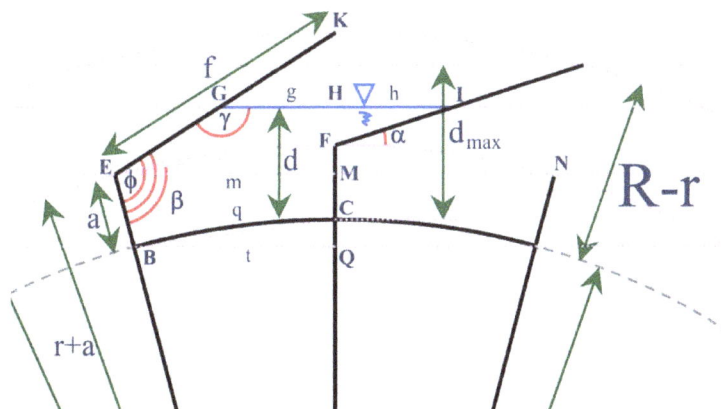

**Figure A1**
**Plank Bucket Design and Fill Variables**

The dominant features of such a design are the use of a radial Rising Board of height a connected at angle $\phi$ to an inclined Float Board of breadth f. This arrangement is of course an attempt to optimise water containment throughout the arc of descent.

In the scheme illustrated above the Bucket Interangle, Denny's $\Delta\theta$ is 15°. Accordingly, the assumed Bucket Count, Denny's n, is 24 ( 360°/15°). R is the Wheel Outer Radius from the center of gyration to the float board edge at Point K, whilst r is the Wheel Inner Radius from the center of gyration to the soal at Points B and C. It is assumed for our simple analyses that the component planks are infinitely thin. d is the Water Fill Depth, which is of course an optimisable feature.

For trial purposes we may assume that the design parameters are: R=12.99, r=10.47, a=1 and d=1.5, whilst n=24; and in the case of Shallow Load we will assume d = 0.3 units.

Derived Parameters

To facilitate analysis we may define the following geometrical parameters:-

$$\Delta\theta = \frac{2\pi}{n}$$
**Equation A2**

$$f = \sqrt{R^2 + (r+a)^2 - 2R(r+a)\cos(\Delta\theta)}$$
**Equation A3**

$$\phi = \pi - \arcsin\left(\frac{R}{f}\sin(\Delta\theta)\right)$$
**Equation A4**

$$\alpha = \phi - \frac{\pi}{2}$$
**Equation A5**

$$\zeta = \Delta\theta + \alpha$$
**Equation A6**

In the above, $\alpha$ is the Angle of Float Board Inclination to the tangent and $\zeta$ is the Board Displacement Angle.

Composite Design Constants

In the following analyses it will prove convenient to help summarise profile equations using the following composite expressions that are constant for any given physical wheel design geometry:-

$$\Xi_1 = r.f.\cos(\zeta).[1 - \cos(\Delta\theta)] + r.\sin(\Delta\theta).[r.(\cos(\Delta\theta)-1)] + a.\cos(\Delta\theta)$$
**Equation A7**

$$\Xi_2 = r^2.\sin(\Delta\theta)$$
**Equation A8**

$$\Xi_3 = \frac{1}{2} \times \left( f.\cos(\zeta)(R - r.\cos(\Delta\theta)) + r.\sin(\Delta\theta)(\cos(\Delta\theta)(r + a) - r) \right)$$
**Equation A9**

## Length Computations

The following auxiliary lengths may further be defined:-

$$m = f\cos(\zeta)$$
**Equation A10**

$$h = \frac{d - a}{\tan\alpha}$$
**Equation A11**

$$g = \frac{R - r - d}{\tan(\zeta)}$$
**Equation A12**

$$v = (r + a)\cos(\alpha)$$
**Equation A13**

$$p = v - r$$
**Equation A14**

$$e = \frac{d - p}{\sin(\zeta)}$$
**Equation A15**

$$q = \sqrt{2r^2(1 - \cos(\Delta\theta))}$$
**Equation A16**

$$t = r \sin \Delta\theta$$
**Equation A17**

$$u = r \cos \Delta\theta$$
**Equation A18**

$$w = r + d$$
**Equation A19**

Area Computations

By reference to Figure A1 it is clear that the wetted and bucket profiles both constitute irregular polygons, except that the soal presents as an arc on the floor of each fill, and therefore descriptive polygons require to be modified by deducting the segment area due to that soal element.

It is convenient to use Meister's Rule for the Area of a General Polygon in order to compute polygonal profiles. Meister's Rule is given by:-

$$A = \frac{1}{2}\left(\sum_{i=1}^{n-1}(x_i.y_{i+1} - x_{i+1}.y_i) + (x_n.y_1 - x_1.y_n)\right)$$
**Equation A20**

Area A should be adjusted by a small decrement due to the soal segment area, $A_{seg}$, which is given by:-

$$A_{seg} = \frac{r^2}{2}(\Delta\theta - \sin(\Delta\theta))$$
**Equation A21**

Therefore, a Corrected Area, $A_{corr}$, is given by:-

$$A = \frac{1}{2}\left(\sum_{i=1}^{n-1}(x_i.y_{i+1} - x_{i+1}.y_i) + (x_n.y_1 - x_1.y_n)\right) - \frac{r^2}{2}(\Delta\theta - \sin(\Delta\theta))$$
**Equation A22**

Water Load Profile Area

The Water Load Profile is of course a function of Depth d and for deeper loads can be resolved into two polygons:-

(a)     Left-of-Center Polygon
    This is a five-sided polygon C(F)HGEB for depths greater than MC which degenerates into a quadrilateral at shallower depths.
(b)     Right-of-Center Polygon

This is a triangular profile extension to CHGEB active for depths that exceed the rising board height a. It is defined by area FIH.

In a case where the water poured into the bucket is too shallow to meet the downstream float board we can, further to the Left-of-Center case, define:-

(c)     Shallow Load Polygon
This is the quadrilateral profile CM′E′B where M′ and E′ are rising board points that are respectively short of M and E, and are at corners of the water surface.

Left of Center Load Profile

The trigonometry of polygons CHGEB and FIH bounding vertices is defined by the following terms matrixes:-

| Point | x Formula | y Formula | Meister's Rule Term |
|---|---|---|---|
| C | $0$ | $r$ | $0 \times (r+a) - 0 \times r$ |
| F | $0$ | $r+a$ | $0 \times (r+d) - 0 \times (r+a)$ |
| H | $0$ | $r+d$ | $0 \times (r+d) - -\dfrac{R-d}{\tan(\zeta)} \times (r+d)$ |
| G | $-\dfrac{R-d}{\tan(\zeta)}$ | $r+d$ | $-\dfrac{R-d}{\tan(\zeta)} \times (r+a)\cos(\Delta\theta) - -f\cos(\zeta) \times (r+d)$ |
| E | $-f\cos(\zeta)$ | $(r+a)\cos(\Delta\theta)$ | $-f\cos(\zeta) \times r\cos(\Delta\theta) - -r\sin(\Delta\theta) \times (r+a)\cos\Delta\theta$ |
| B | $-r\sin(\Delta\theta)$ | $r\cos(\Delta\theta)$ | $-r\sin\Delta\theta \times r - 0 \times r\cos\Delta\theta$ |
| F | $0$ | $r+a$ | $0 \times (r+d) - \dfrac{d-a}{\tan(\alpha)} \times (r+a)$ |
| I | $+\dfrac{d-a}{\tan(\alpha)}$ | $r+d$ | $(r+d) \times \dfrac{d-a}{\tan(\alpha)} - 0 \times (r+d)$ |
| H | $0$ | $r+d$ | $0 \times (r+a) - 0 \times (r+d)$ |

**Table A1**
**Terms Matrix for Left and Right of Center Bucket Load Profile**
**Trigonometric Expressions**

| Point | x | y | x Formula | y Formula |
|-------|-----|--------|-----------|-----------|
| C | 0 | 10.47 | 0 | $r$ |
| F | 0 | 11.47 | 0 | $r + a$ |
| H | 0 | 11.97 | 0 | $r + d$ |
| G | -1.584665 | 11.97 | $-\dfrac{R - r - d}{\tan(\zeta)}$ | $r + d$ |
| E | -2.968654 | 11.079169 | $-f\cos(\zeta)$ | $(r + a)\cos(\Delta\theta)$ |
| B | -2.709835 | 10.113243 | $-r\sin(\Delta\theta)$ | $r\cos(\Delta\theta)$ |
| F | 0 | 11.47 | 0 | $r + a$ |
| I | 1.560299 | 11.97 | $+\dfrac{d - a}{\tan(\alpha)}$ | $r + d$ |
| H | 0 | 11.97 | 0 | $r + d$ |

**Table A2**
**Terms Matrix for Left and Right of Center Bucket Load Profile**
**Approximate Vertex Values**

The Meister's Rule expression for Left-of-Center $A_{loadl}$ is accordingly:-

$$A_{loadl} = \frac{1}{2} \times \begin{bmatrix} \left(0 \times (r+a) - 0 \times r\right) + \left(0 \times (r+d) - 0 \times (r+a)\right) + \left(0 \times (r+d) - -\dfrac{R-r-d}{\tan(\zeta)} \times (r+d)\right) + \\ \left(-\dfrac{R-r-d}{\tan(\zeta)} \times (r+a)\cos(\Delta\theta) - -f\cos(\zeta) \times (r+d)\right) + \\ \left(-f\cos(\zeta) \times r\cos(\Delta\theta) - -r\sin(\Delta\theta) \times (r+a)\cos\Delta\theta\right) + \left(-r\sin\Delta\theta \times r - 0 \times r\cos\Delta\theta\right) \end{bmatrix}$$

**Equation A23**

After we have eliminated null terms we may note that $\cos(\zeta)/\sin(\zeta)$ is a common factor in the fourteen terms that remain. Noting further that $\cos(\zeta)/\sin(\zeta) = \cot(\zeta)$, we may regroup and rearrange Equation A23 to give:-

$$A_{loadl} = \frac{1}{2} \times \cot(\zeta) \times \begin{bmatrix} (R-r)(r+d) - d(r+d) - \cos(\Delta\phi).[(R-r)(r+a) - d(r+a)] \\ + f.\cos(\zeta).[r.(1 - \cos(\Delta\theta)) + d] + r.\sin(\Delta\theta).[r(\cos(\Delta\theta) - 1) + a.\cos(\Delta\theta)] \end{bmatrix}$$

**Equation A24**

At this juncture it is convenient to define the Radial Depth Function, $\Phi(x,y,z,u,v)$, as:-

$$\Phi(x, y, z, u, v) = (x - y) \times (y + z) - u \times (v + y)$$

**Equation A25**

It is now possible to write a shortened but uncompromised $A_{loadl}$ as:-

$$A_{loadl} = \frac{1}{2} \times \cot(\zeta) \times \left[ \Phi(R,r,d,d,d) - \cos(\Delta\theta).\Phi(R,r,a,d,a) + df.\cos(\zeta) + \Xi_1 \right]$$

**Equation A26**

Using the stated data, Equation A26 yields about 4.287243 square units, but this must of course be corrected for the soal segment, which in this case is around 0.163354 square units. Accordingly, the corrected Left-Hand Load Area, $A_{loadlcorr}$, is 4.123889 square units.

Right of Center Load Profile

There are allegedly several thousand alternative formulae for the area of a triangle, several of which cater to the right-angled case: We, however, will stick with Meister's Rule. In terms of that the Right-Hand Load Area, $A_{loadr}$, is specified by:-

$$A_{loadr} = \frac{1}{2} \times \left[ \left( 0 \times (r+d) - \frac{d-a}{\tan(\alpha)} \times (r+a) \right) + \left( \frac{d-a}{\tan(\alpha)} \times (r+d) - 0 \times (r+d) \right) + \left( 0 \times (r+a) - 0 \times (r+d) \right) \right]$$

**Equation A27**

that soon reduces to:-

$$A_{loadr} = \frac{1}{2}.\cos(\alpha).\frac{d^2 - 2da + a^2}{\sin(\alpha)} = \frac{1}{2}.\cot(\alpha).(d-a)^2$$

**Equation A28**

Depth does Not Meet the Float Board

When the fill water is too shallow to reach the downstream float board we may adapt our Left of Center treatment and of course omit the Right of Center supplement.
Consider a depth on MC. That is a depth for the condition $d_{crit} > d > 0$ where:-

$$d_{crit} = (r + a).\cos(\Delta\theta) - r = r(\cos(\Delta\theta) - 1) + a\cos(\Delta\theta)$$

**Equation A29**

For trial purposes, we will take it that $d = 0.3$ units for this analysis.
The trigonometry of the relevant water profile quadrilateral CM′E′B is given by the following terms matrices:-

| Point | x Formula | y Formula | Meister's Rule Term |
|---|---|---|---|
| C | 0 | $r$ | $0 \times (r+d) - 0 \times r$ |
| M′ | 0 | $r+d$ | $0 \times (r+d) - (r+d) \times (r+d)\tan(\Delta\theta)$ |
| E′ | $-(r+d)\tan(\Delta\theta)$ | $r+d$ | $-(r+d)\tan(\Delta\theta) \times r\cos(\Delta\theta) - -r\sin(\Delta\theta) \times (r+d)$ |
| B | $-r\sin(\Delta\theta)$ | $r\cos(\Delta\theta)$ | $r \times -r\sin(\Delta\theta) - 0 \times r\cos(\Delta\theta)$ |

**Table A3**
**Terms Matrix for the Shallow Bucket Load Profile**
**Trigonometric Expressions**

| Point | x | y | x Formula | y Formula |
|---|---|---|---|---|
| C | 0 | 10.47 | 0 | $r$ |
| M′ | 0 | 10.77 | 0 | $r+d$ |
| E′ | -2.885813 | 10.77 | $-(r+d)\tan(\Delta\theta)$ | $r+d$ |
| B | -2.709835 | 10.113243 | $-r\sin(\Delta\theta)$ | $r\cos(\Delta\theta)$ |

**Table A4**
**Terms Matrix for the Shallow Bucket Load Profile**
**Approximate Vertex Values**

The Meister's Rule expression for the Shallow Load Polygon A<sub>loads</sub> is accordingly:-

$$A_{loads} = \frac{1}{2} \times \begin{bmatrix} (0 \times (r+d) - 0 \times r) + \\ (0 \times (r+d) - (r+d) \times -(r+d)\tan(\Delta\theta)) + \\ (-(r+d)\tan(\Delta\theta) \times r\cos(\Delta\theta) - -r\sin(\Delta\theta) \times (r+d)) + \\ (r \times -r\sin(\Delta\theta) - 0 \times r\cos(\Delta\theta)) \end{bmatrix}$$

**Equation A30**

Noting that the third summand is zero and that further zero terms may be eliminated we may regroup and rearrange Equation A30 to give:-

$$A_{loads} = \frac{1}{2} \times \left( \tan(\Delta\theta).(r+d)^2 - r^2.\sin(\Delta\theta) \right)$$

**Equation A31**

or:-

$$A_{loads} = \frac{1}{2} \times \left( \tan(\Delta\theta).(r+d)^2 - \Xi_3 \right)$$

**Equation A32**

Consolidated Water Profile Area

Noting that Shallow Load Critical Depth d<sub>crit</sub> is given by Equation A29 we may express the Total Water Profile Area, A<sub>load</sub>, as:-

$$A_{load} = if\left(d < d_{crit}, A_{loads}, if\left(d < a, A_{loadl}, A_{loadl} + A_{loadr}\right)\right) - A_{seg}$$

**Equation A33**

## The Accommodation Area

In Figure A1 the Accommodation Area is defined by the quadrilateral C(F)(H)KEB, which must be suitably decremented by the Soal Segment Area, $A_{seg}$, before use in containment studies.

The Accommodation Area is in effect the maximum possible capacity of a bucket when it is turned to $\pi/2$ radians beyond Top Center.

The Accommodation Area is defined by the following terms matrices:-

| Point | x Formula | y Formula | Meister's Rule Term |
|-------|-----------|-----------|---------------------|
| C | $-r\sin(\Delta\theta)$ | $r\cos(\Delta\theta)$ | $-r\sin(\Delta\theta)\times r - 0\times r\cos(\Delta\theta)$ |
| M′ | $0$ | $r$ | $0\times R - 0\times r$ |
| E′ | $0$ | $R$ | $0\times(r+a)\cos(\Delta\theta) - -f\cos(\zeta)\times R$ |
| B | $-f\cos(\zeta)$ | $(r+a)\cos(\Delta\theta)$ | $-f\cos(\zeta)\times r\cos(\Delta\theta) - -r\sin(\Delta\theta)\times(r+a)\cos(\Delta\theta)$ |

**Table A5**
**Terms Matrix for the Accommodation Area Profile**
**Trigonometric Expressions**

| Point | x | y | x Formula | y Formula |
|-------|-----|-----|-----------|-----------|
| C | -2.709835 | 10.113243 | $-r\sin(\Delta\theta)$ | $r\cos(\Delta\theta)$ |
| M′ | 0 | 10.47 | $0$ | $r$ |
| E′ | 0 | 12.99 | $0$ | $R$ |
| B | -2.968654 | 11.079169 | $-f\cos(\zeta)$ | $(r+a)\cos(\Delta\theta)$ |

**Table A6**
**Terms Matrix for the Accommodation Area Profile**
**Approximate Vertex Values**

The Meister's Rule expression for the Accommodation Area, $A_{ac}$, is accordingly:-

$$A_{ac} = \frac{1}{2}\times \begin{bmatrix} \left(-r\sin(\Delta\theta)\times r - 0\times r\cos(\Delta\theta)\right)+ \\ \left(0\times R - 0\times r\right)+ \\ \left(0\times(r+a)\cos(\Delta\theta) - -f\cos(\zeta)\times R\right)+ \\ \left(-f\cos(\zeta)\times r\cos(\Delta\theta) - -r\sin(\Delta\theta)\times(r+a)\cos(\Delta\theta)\right) \end{bmatrix}$$

**Equation A34**

On the immediate elimination of zero terms this reduces to:-

$$A_{ac} = \frac{1}{2} \times \begin{bmatrix} (-r\sin(\Delta\theta) \times r) + \\ (f\cos(\zeta) \times R) + \\ (-f\cos(\zeta) \times r\cos(\Delta\theta) - -r\sin(\Delta\theta) \times (r+a)\cos(\Delta\theta)) \end{bmatrix}$$

**Equation A35**

Further simplification allows us to write:-

$$A_{ac} = \frac{1}{2} \times [(f\cos(\zeta)(R - r\cos(\Delta\theta))) + r.\sin(\Delta\theta).(\cos(\Delta\theta)(r+a) - r)]$$

**Equation A36**

or:-

$$A_{ac} = \Xi_3$$

**Equation A37**

This should be adjusted by $A_{seg}$ as:-

$$A_{ac} = \Xi_3 - A_{seg}$$

**Equation A38**

## Calculation of the Accommodation Defect Area for the Determination of $\theta_{spill}$

We are finally in a position to compute the Accommodation Defect Area, $A_{def}$, that will enable us to move forward to calculate $\theta_{spill}$, the Intrabucket Water Surface Angle of Incipient Spill ( effectively Denny's $\theta_1$ ), and to pass forward $A_{load}$ as the progressively-decremented bucket water load for power and efficiency calculations.

The required Defect Area is given by:-

$$A_{def} = A_{ac} - A_{load}$$

**Equation A39**

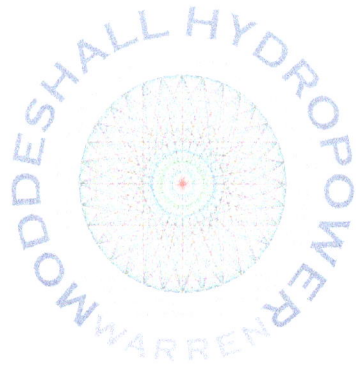

## TECHNICAL APPENDIX B
## ANALYSES OF PROGRESSIVE BUCKET EMPTYING

In order to prosecute analysis of the power and efficiency of overshot roman wheels during the half-cycle in which water inhabits buckets we will need carefully to examine the trigonometry of two key system properties.

Firstly, the Area Defect, that is the area of the bucket profile that is unwetted at the point when the bucket contents is just about to start leaving the container. This will develop into a triangular area of the profile modified by the arc of the soal segment at its inner boundary. This area is a compliment of $A_{load}$, but its triangularity, and the critical angle it controls, is modified by the soal arc.

Secondly, the Critical Angle of Incipient Spill, $\theta_{spill}$, is directly related to $A_{def}$, and other angles $\theta$ may be computed to define the progress of the residual bucket contents as the bucket processes along its half-cycle. I must emphasise straight away that $\theta_{spill}$ and $\theta$ must not be confused with Denny's $\theta$ and $\Delta\theta$, to which they have no direct relation that I can ascertain.

Figure B1 shows the salient geometrical and trigonometrical factors bearing upon the wheel profile and schematises an incipiently-spilling bucket at bottom left.

In order to clarify the spacial relationships the spilling bucket is, conceptually speaking, rotated dextrally through about 210° and the relevant diagram cropped and enlarged in Figure B2.

Triangle boxed AT ( i.e. BFC ) approximates the compliment of $A_{load}$ and its angle at B, $\theta$, is near to $\theta_{spill}$, simple correctors being required to gain exactitude. The triangle boxed AS ( CFD ) and its angle at D, $\chi$, complement AT and $\theta$, to which they have definite algebraic relations. AS and $\chi$ shall assist our mathematical analysis.

Angle $\varepsilon$ ( not illustrated ) is the obtuse angle FCB: It also assists analysis.

R, r, a, $\Delta\theta$, and $\phi$ are all design constants. They are respectively Wheel Outer Radius from DB; Wheel Inner Radius from D to C; a Rising Board Radial Height; Denny's Interbucket Arc Angle ( i.e. $2\pi$/NumberOfBuckets ); and The Internal Angle between a Rising Board and a Float Board.

For present trial purposes we will assume that R=15, r=10, AT=7.28 square units and n=12, therefore $\Delta\theta$ is 30°. Arat, the ratio of triangle areas AS/AT is 2.

For these inputs, the salient outcomes include: $\theta_{spill}$=0.4919896 radians ( 28.1889276° ); $\chi_{spill}$=0.295481 radians; $w_{spill}$=6.1645228 length units.

The rising board is assumed to be 1.85 units high. Under these conditions the water surface profile trace length at the rising board's soal contact, $w_{rising0}$, is 8.074179764 length units; the attendant $\theta$ value, $\theta_{rising0}$ is 0.667797288 radians, and the Apical Angle, $\theta_{max}$ is 0.89630253.

Later, another trial conformation will present different geometrical outcomes.

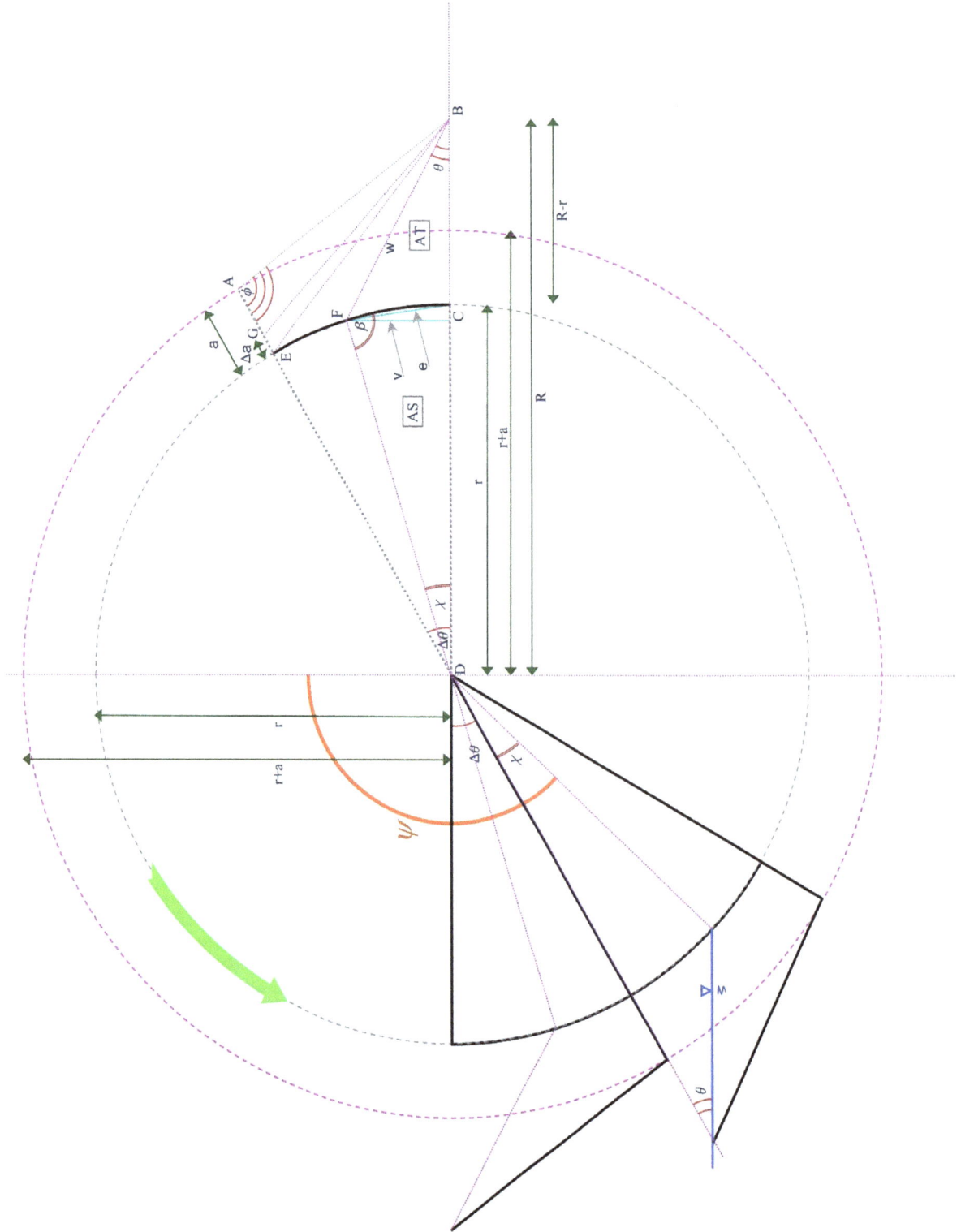

**Figure B1**
**Overshot Waterwheel Geometrical and Trigonometrical Profile**

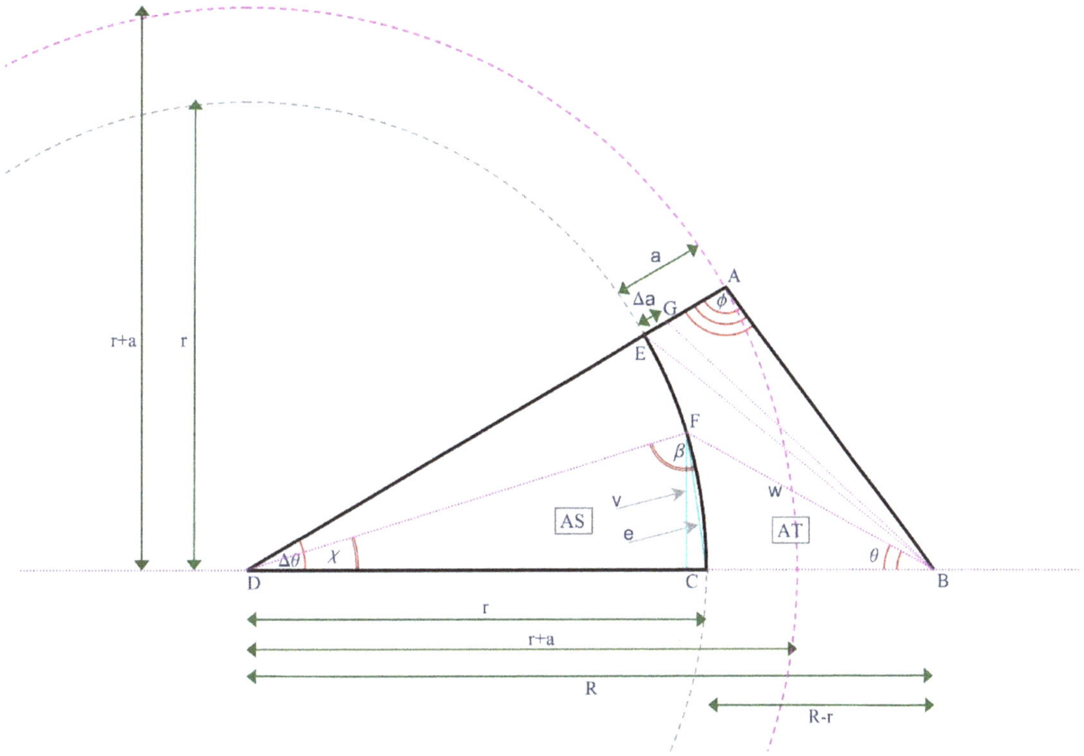

**Figure B2**
**Overshot Waterwheel Bucket Containment Progress**
**Analysis Diagram**

Auxiliary Relations

Useful ancillary relations include:-

$$\beta = \frac{\pi - \chi}{2}$$
**Equation B1**

$$\varepsilon = \pi - \beta = \frac{\pi + \chi}{2}$$

**Equation B2**

$$e = \sqrt{2r^2\left(1 - \cos(\chi)\right)}$$
**Equation B3**

$$v = r\sin(\chi)$$
**Equation B4**

$$w = \sqrt{\left(R^2 - r^2\right) - 2Rr\cos(\chi)}$$

**Equation B5**

$$AS = \frac{1}{2}r^2 \sin(\chi)$$

**Equation B6**

$$AT = \frac{1}{2}e(R - r)\sin(\varepsilon)$$

**Equation B7**

$$Area(AS + AT) = Area\ of\ Triangle\ BFD = \frac{1}{2}Rr\sin(\chi)$$

**Equation B8**

Composite Design Constants

The following composite design constants assist and economise computation:-

$$\Xi_1 = R - r \qquad \Xi_2 = R^2 + r^2 \qquad \Xi_3 = R^2 - r^2$$

**Equation B9**       **Equation B10**       **Equation B11**

$$\Xi_4 = (R - r)^2 \qquad \Xi_5 = r(R - r) \qquad \Xi_6 = 2Rr$$

**Equation B12**       **Equation B13**       **Equation B14**

$$\Xi_7 = \frac{2}{\Xi_1} \qquad \Xi_8 = \frac{8\Xi_2}{\Xi_4} \qquad \Xi_9 = \frac{(4R)^2}{(r - R)^2}$$

**Equation B15**       **Equation B16**       **Equation B17**

$$\Xi_{10} = \Xi_8.f\theta - \Xi_9 \qquad \Xi_{11} = \left[\left(\Xi_7\right)^2.f\theta\right]^2 \qquad \Xi_{12} = \Xi_9 - \Xi_8.f\theta$$

**Equation B18**       **Equation B19**       **Equation B20**

$$\Xi_{13} = -\left[\left(\Xi_7\right)^2.f\theta\right]^2$$

**Equation B21**

The General Definition of Bucket Water Surface Angle, θ

The progressive Bucket Water Surface Angle, θ, is defined by:-

$$\theta = \arcsin\left(\frac{v}{w}\right)$$

**Equation B22**

## AS and AT Triangle Ratio Relations

The $\chi$ and $\theta$ triangles are complementary and their ratio, Arat, has a definite relationship.

Specifically:-

$$Arat = \frac{AS}{AT} = \frac{\frac{1}{2}r^2 \sin(\chi)}{\frac{1}{2}e(R-r)\sin(\varepsilon)} = \frac{\frac{1}{2}r^2 \sin(\chi)}{\frac{1}{2}\sqrt{2r^2(1-\cos(\chi))}(R-r)\sin\left(\frac{\pi+\chi}{2}\right)}$$

$$= \frac{r^2 \sin(\chi)}{\sqrt{2r^2(1-\cos(\chi))}(R-r)\sin\left(\frac{\pi+\chi}{2}\right)}$$

**Equation B23**

Further to simplify Equation B23 it is convenient to square both sides to yield:-

$$Arat^2 = \left(\frac{r^2 \sin(\chi)}{\sqrt{2r^2(1-\cos(\chi))}(R-r)\sin\left(\frac{\pi+\chi}{2}\right)}\right)^2 = \frac{1}{2}(\cos(\chi)-1)\frac{r^2}{\left(R^2-2Rr+r^2\right)\cos\left(\frac{1}{2}\chi\right)^2}$$

**Equation B24**

Rearrangement then gives:-

$$\frac{\frac{1}{2}(\cos(\chi)-1)}{\cos\left(\frac{1}{2}\chi\right)^2} = \frac{Arat^2}{r^2}\left(R^2-2Rr+r^2\right) = Arat^2\left(\frac{R^2}{r^2}-\frac{2R}{r}+1\right)$$

**Equation B25**

Noting that the LHS of Equation B25 is unity we may rearrange the RHS for Arat, and then simplify:-

$$Arat = \sqrt{\frac{1}{\frac{R^2}{r^2}-\frac{2R}{r}+1}} = \sqrt{\frac{r^2}{R^2-2Rr+r^2}} = \sqrt{\frac{r^2}{(R-r)(R-r)}} = \sqrt{\frac{r^2}{(R-r)^2}} = \frac{r}{R-r}$$

**Equation B26**

Therefore the triangle areas AS and AR are in the simple ratio r/(R-r) at all points along the $\Delta\theta$ arc CE.

## The Expression of Bucket Water Surface Angle, θ , in terms of Area AT

We may now move forward to link the uncorrected progressive Water Surface Angle θ with the uncorrected triangle area AT. By "uncorrected" we mean of course uncorrected for the progressive Soal Segment Area, $A_{seg}$, which will have the net effect of slightly reducing actual θ for any given $A_{def}$. Put another way, AT slightly overstates $A_{def}$ and its determination of θ.

From the forgoing we now know that:-

$$\frac{AS}{AT} = \frac{\frac{1}{2}r^2 \sin(\chi)}{AT} = \frac{r}{R-r}$$

**Equation B27**

Rearrangement shows that:-

$$\frac{1}{2}r^2 \sin(\chi) = \frac{r.AT}{R-r}$$

$$\therefore \sin(\chi) = \frac{2.r.AT}{r^2(R-r)}$$

$$\therefore \sin(\chi) = \frac{2.AT}{r(R-r)}$$

$$\therefore \chi = \arcsin\left(\frac{2.AT}{r(R-r)}\right)$$

**Equation B28**

Using the definition of θ we may now make the appropriate substitution to express θ in terms of AT ( supplied in terms of $A_{def}$ ), and the design constants:-

$$\theta = \arcsin\left(\frac{v}{w}\right) = \arcsin\left(\frac{r\sin\left(\arcsin\left(\frac{2.AT}{r(R-r)}\right)\right)}{\sqrt{R^2 + r^2 - 2Rr\cos\left(\arcsin\left(\frac{2.AT}{r(R-r)}\right)\right)}}\right)$$

**Equation B29**

By eliminating the redundant counterfunctionalities we may re-express this as:-

$$\theta = \arcsin\left(\frac{r\left(-2\dfrac{AT}{r(r-R)}\right)}{\sqrt{R^2 + r^2 - 2Rr\dfrac{\sqrt{r^2(R-r)^2 - 4.AT^2}}{r(R-r)}}}\right)$$

**Equation B30**

Further simplifications and re-arrangements bring us to:-

$$\theta = \arcsin\left(\frac{1}{\dfrac{R-r}{2.AT}\sqrt{R^2 + r^2 - 2Rr\dfrac{\sqrt{r^2(R-r)^2 - 4.AT^2}}{r(R-r)}}}\right)$$

**Equation B31**

Or for greater solutional stability:-

$$\theta = \arcsin\left(\frac{\dfrac{2.AT}{R-r}}{\sqrt{R^2 + r^2 - 2Rr\dfrac{\sqrt{r^2(R-r)^2 - 4.AT^2}}{r(R-r)}}}\right)$$

**Equation B32**

Noting the Design Constants $\Xi_i$ we may conveniently cast this Equation B32 as:-

$$\theta = \arcsin\left(\frac{\dfrac{2.AT}{\Xi_1}}{\sqrt{\Xi_2 - \Xi_6\dfrac{\sqrt{(\Xi_5)^2 - 4.AT^2}}{\Xi_5}}}\right)$$

**Equation B33**

Further economy can of course be incorporated by noting that $4.AT^2$ is the square of $2.AT$.

## Transposition for AT

To identify the Unadjusted Water Profile AT attaching to any particular Surface Angle $\theta$ we may transpose Equation B32 to obtain the quadratic expression:-

$$-\left(\frac{(\Xi_7)^2}{\sin^2(\theta)}\right)^2 \cdot (AT^2)^2 + \left(2\Xi_2 \frac{(\Xi_7)^2}{\sin^2(\theta)} - \frac{4.(\Xi_6)^2}{(\Xi_5)^2}\right) \cdot AT^2 + \left((\Xi_6)^2 - \Xi_2{}^2\right) = 0$$

**Equation B34**

By applying The Equation for the Roots of a Quadratic Equation we obtain the two real roots for $AT^2$:-

$$Rt(AT^2) = \frac{\left[-\left[2(R^2+r^2)\frac{\left(\frac{2}{R-r}\right)^2}{\sin^2(\theta)} - \frac{(2Rr)^2}{[r(R-r)]^2}\right] \pm \sqrt{\left[2(R^2+r^2)\frac{\left(\frac{2}{R-r}\right)^2}{\sin^2(\theta)} - \frac{(2Rr)^2}{[r(R-r)]^2}\right]^2 - 4.\left[\frac{\left(\frac{2}{R-r}\right)^2}{\sin^2(\theta)}\right]^2 \left[(2Rr)^2 - (R^2-r^2)^2\right]}\right]}{2.\left[\frac{\left(\frac{2}{R-r}\right)^2}{\sin^2(\theta)}\right]^2}$$

**Equation B35**

Because only the root involving the positive square root term is of interest to us we may adopt:-

$$Rt(AT^2) = \frac{\left[-\left[2(R^2+r^2)\frac{\left(\frac{2}{R-r}\right)^2}{\sin^2(\theta)} - \frac{(2Rr)^2}{[r(R-r)]^2}\right] + \sqrt{\left[2(R^2+r^2)\frac{\left(\frac{2}{R-r}\right)^2}{\sin^2(\theta)} - \frac{(2Rr)^2}{[r(R-r)]^2}\right]^2 - 4.\left[\frac{\left(\frac{2}{R-r}\right)^2}{\sin^2(\theta)}\right]^2 \left[(2Rr)^2 - (R^2-r^2)^2\right]}\right]}{2.\left[\frac{\left(\frac{2}{R-r}\right)^2}{\sin^2(\theta)}\right]^2}$$

**Equation B36**

This may conveniently be re-arranged to isolate the Function of Theta, fθ, from functions of design constants only in the following terms:-

$$Rt(AT^2) = \frac{\left[-\left[\frac{8(R^2+r^2)}{(R-r)^2}\frac{1}{\sin^2(\theta)} - \frac{(4R)^2}{(r-R)^2}\right] + \sqrt{\left[\frac{8(R^2+r^2)}{(R-r)^2}\frac{1}{\sin^2(\theta)} - \frac{(4R)^2}{(r-R)^2}\right]^2 - 4.\left[\left(\frac{2}{R-r}\right)^2\frac{1}{\sin^2(\theta)}\right]^2 \left[(R^2+r^2)^2 - (2Rr)^2\right]}\right]}{2.\left[\left(\frac{2}{R-r}\right)^2\frac{1}{\sin^2(\theta)}\right]^2}$$

**Equation B37**

If we now define the Computational Theta Function, fθ, as:-

$$f\theta = \frac{1}{\sin^2\theta}$$

**Equation B38**

we may now write a partial rationalisation of the algebraic structure as:-

$$Rt\left(AT^2\right) = \frac{(\Xi_9 - \Xi_8.f\theta) + \sqrt{(\Xi_9 - \Xi_8.f\theta)^2 - 4.-\left[(\Xi_7)^2.f\theta\right]^2.\left[(\Xi_2)^2 - (\Xi_6)^2\right]}}{2.-\left[(\Xi_7)^2.f\theta\right]^2}$$

**Equation B41**

Clearly, the Unadjusted Unwetted Water Profile Area, AT, is now given by:-

$$AT = \sqrt{Rt\left(AT^2\right)}$$

**Equation B42**

We may note that this value for AT now permits a convenient access to Water Surface Profile ray w and to Gyratory Co-Angle, $\chi$, in terms of:-

$$w = \frac{2.AT}{(R - r).\sin(\theta)}$$

**Equation B43**

$$\chi = \arccos\left(\frac{w^2 - R^2 - r^2}{-2Rr}\right)$$

**Equation B44**

Corrections for the Soal Segment Area

To employ AT it is required to modify the raw triangular area by the attaching Soal Segment Area, $A_{seg}$, to give a Corrected Unwetted Profile Area, $A_{corr}$. This is done using:-

$$A_{corr} = AT - A_{seg} = AT - \frac{r^2}{2}(\chi - \sin\chi)$$

**Equation B45**

where $\chi$ is the raw value of $\chi$ as used in the computation of AT.
An adjusted value of $\chi$ is now given by:-

$$\chi_{corr} = \arcsin\left(\frac{2 \times AT_{corr}}{r(R - r)}\right)$$

**Equation B46**

If it is required to know the corrected Progress Angle $\psi$ corresponding to $AT_{corr}$, this is given by:

$$\psi = \frac{\pi}{2} + \theta + \chi_{corr}$$

**Equation B47**

## Area-Theta Relations in the Rising Board Contact Regime

When the bucket water surface profile touches the base of the rising board at the downstream corner of the wheel bucket it leaves the geometrical regime that we have discussed at length above. The ratio of triangles AS and AT is no longer relevant as, mathematically-speaking, their growth has ceased.

The surface profile trace, w, now touches the base of the rising board and begins to creep along it to the float board, and eventual complete tip-out as the progress of the wheel turning continues.

Likewise, further corrections for the soal segment area are not required, because the developing triangle of unwetment now has its base entirely upon the rising board.

With reference to Figure B2 our attention is now focused upon the trigonometry of triangle BAE.

The water surface trace, $w_{rising0}$, at the contact of the soal and the rising board is given by:-

$$w_{ri \sin g0} = \sqrt{R^2 + r^2 - 2Rr\cos(\Delta\theta)}$$

**Equation B48**

and the corresponding value of $\theta$ at that point, $\theta_{rising0}$, is given by:-

$$\theta_{ri \sin g0} = \arcsin\left(\frac{r}{w_{ri \sin g0}}\sin(\Delta\theta)\right)$$

**Equation B49**

The limiting value of $\theta$, Apical Angle $\theta_{max}$, at the float board is:-

$$\theta_{max} = \arcsin\left(\frac{r+a}{f}\sin(\Delta\theta)\right)$$

**Equation B50**

In order to compute the progressive decrements of wetted area against the rising board it is convenient to do so with regard to Increment of Rising Board Height, $\Delta a$, which is some fraction of the length a.

First compute local Water Surface Profile Length, $w_{rising}$, at $\Delta a$ using:-

$$w_{ri \sin g} = \sqrt{R^2 + (r + \Delta a)^2 - 2R(r + \Delta a)\cos(\Delta\theta)}$$

**Equation B51**

then the corresponding local $\theta$, $\theta_{rising}$, by:-

$$\theta_{ri\sin g} = \arcsin\left[\frac{r + \Delta a}{w_{ri\sin g}}\sin(\Delta\theta)\right]$$

**Equation B52**

From which we may establish the relevant Area Decrement, A$_{rising}$, as:-

$$A_{ri\sin g} = \frac{1}{2}.w_{ri\sin g0}.w_{ri\sin g}.\sin(\theta_{ri\sin g} - \theta_{ri\sin g0})$$

**Equation B53**

In any perceptible spacial sense $\chi$ does not appear to relate to area relations along the rising board, because, like AS, we seem to have left it behind with the soal. Nevertheless, $\chi$ continues steadily to decline, as it were in hyperspace, and can be computed, and employed in computing the equally coevolving Wheel Progress Angle, $\psi$.

At the rising board, $\chi_{rising}$, is given by:-

$$\chi_{ri\sin g} = \arccos\left[\frac{w_{ri\sin g}^2 - (R^2 + r^2)}{2Rr}\right]$$

**Equation 54**

The Wheel Progress Angle, $\psi_{rising}$, is now given by:-

$$\psi_{ri\sin g} = \frac{\pi}{2} + \chi_{ri\sin g} + \theta_{ri\sin g}$$

**Equation B55**

In progressive terms, the Unwetted Area, A$_{u,i}$, can be related to the Wetted Area, A$_{wet,i}$, using the logic:-

$$A_{wet,i} = if\left(A_{u,i} < A_{def}, A_{load}, A_{wet,i-1} - A_{u,i} + A_{u,i-1}\right)$$

**Equation B56**

Some Curious Algebraic Fitments

It is notable that several geometrical outcomes of overshot roman wheel geometry have simple algebraic polynomial relations to generative variables such as $\theta$.

Note that all coefficients are special to particular sets of design parameters.

$\chi$ versus Triangle Areas AS and AT

The Co-Angle $\chi$ very nearly but not quite grows linearly with AS and AT for most design configurations as is shown qualitatively below:-

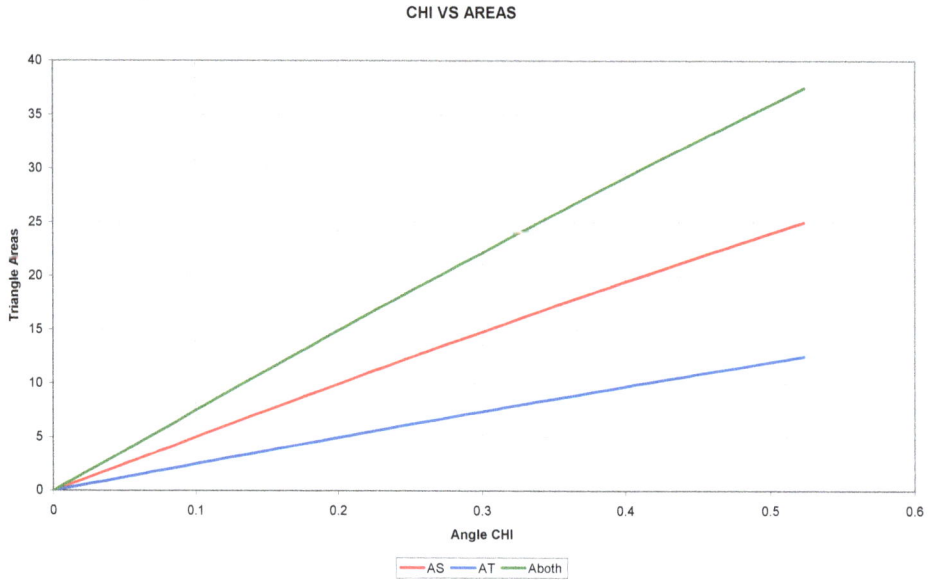

**CHI VS AREAS**

**Figure B3**
**Quasi-Linearity of χ versus Triangle Areas AT and AS**

Soal Segment Basal Lines e and v

e and v are nearly the same even for low-r configurations with pronounced soal-arcs. When R is slightly longer than r in tall shallow-bucket wheel designs, e and v become nearly identical:-

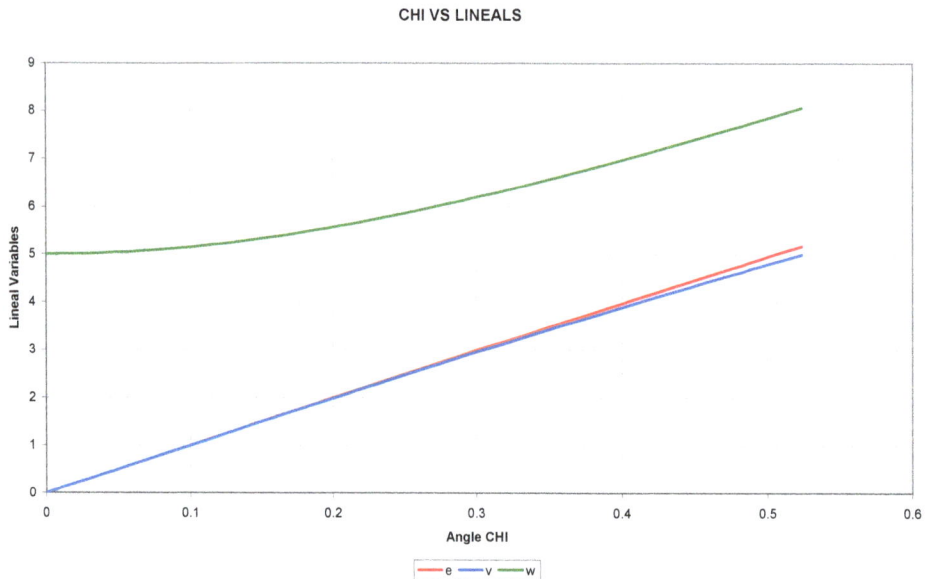

**CHI VS LINEALS**

**Figure B4**
**Soal Basal Lines e and v are Nearly the Same**

Corrected AT Triangle Area is Cubic with θ

For at least several design configurations $AT_{corr}$ varies with θ according to a cubic algebraic polynomial:-

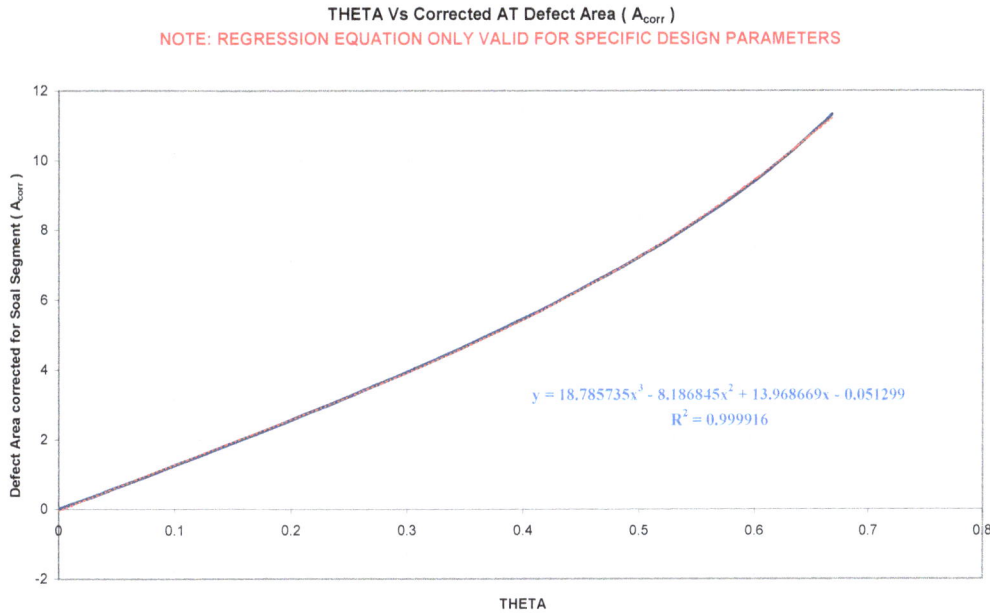

**THETA Vs Corrected AT Defect Area ( $A_{corr}$ )**
NOTE: REGRESSION EQUATION ONLY VALID FOR SPECIFIC DESIGN PARAMETERS

$y = 18.785735x^3 - 8.186845x^2 + 13.968669x - 0.051299$
$R^2 = 0.999916$

**Figure B5**
**Cubic Variation of $A_{corr}$ versus θ**

Rising Board: w Varies Quadratically with θ

For the Rising Board sweep regime, the behaviour of the Water Surface Profile Trace, w, varies as a second-degree algebraic polynomial with Sweep Angle, θ:-

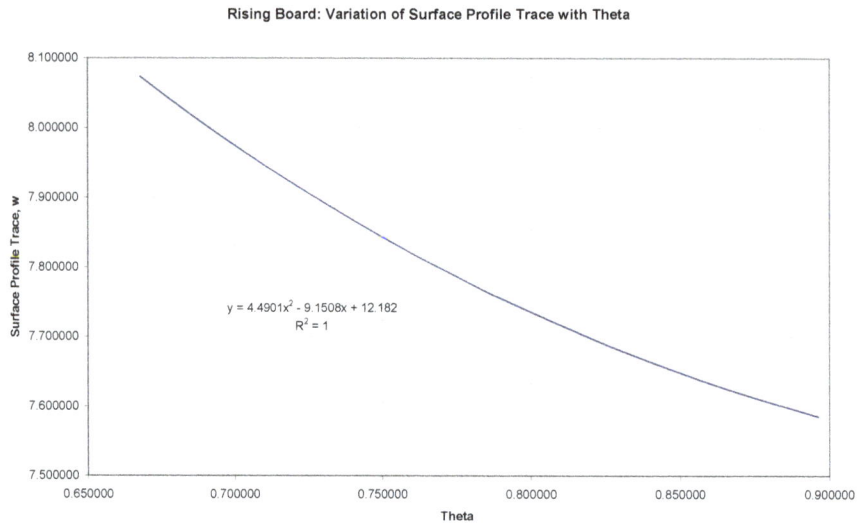

Rising Board: Variation of Surface Profile Trace with Theta

$$y = 4.4901x^2 - 9.1508x + 12.182$$
$$R^2 = 1$$

**Figure B6**
**Quadratic Variation of $w_{rising}$ with $\theta_{rising}$**

Rising Board: Swept Area is Linear with $\theta$

It is notable that Equation B52 is a linear relation between $\theta_{rising}$ and $A_{rising}$, but it is possibly not worthwhile to provide a linear regression, as it would be special to whatever design parameters were chosen. Nevertheless, I include a plot of $\theta_{rising}$ versus $A_{rising}$ in Figure B7 for demonstration purposes:-

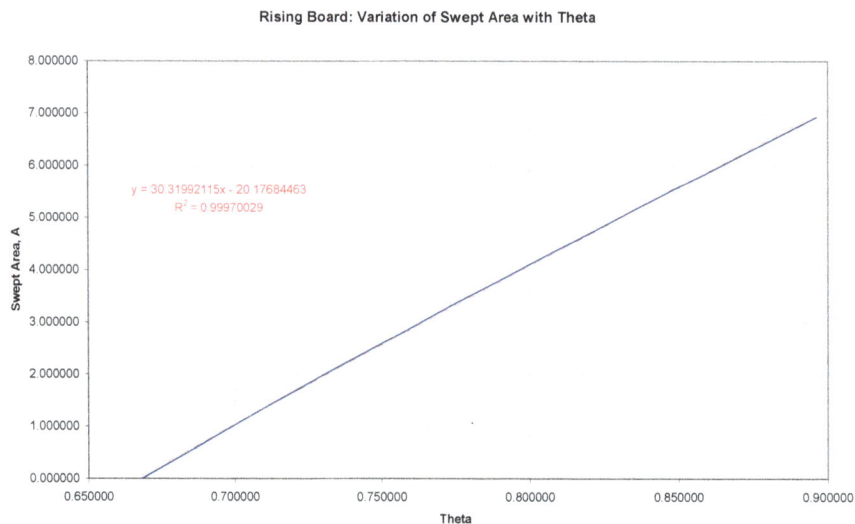

Rising Board: Variation of Swept Area with Theta

$$y = 30.31992115x - 20.17684463$$
$$R^2 = 0.99970029$$

**Figure B7**
**The Essential Linearity of Rising Board**
**Unwetted Area with Sweep Angle $\theta_{rising}$**

Further research is needed to determine whether these polynomials have any larger scientific or computational value.

## Profile Geometry as a Metric of Hydraulic Efficiency

A one-hundred percent efficient overshot waterwheel is one of the classic components of a famous perpetual motion paradigm: It is of course impossible.

Of interest to us is the development, and in practice the diminution, of the bucket water profile as it processes through the $\pi$ radian arc from Top Center ( TC ) to Bottom Center ( BC ). Because a bucket is of constant lateral section, and identical to others, that profile is a linear analog of contained potential energy.

Explicitly:-

$$E = mgh = gh.\rho V = gh.\rho AW = \Theta A$$

**Equation B57**

where E is local Potential Energy of the full or part-full bucket; m is Contained Water Mass; g is the regional Acceleration Due to Gravity; h is Height above Datum; $\rho$ is the temporal Density of Water; A is the local Water Profile Area; W is the design Bucket Width; and $\Theta$ is a Compound Constant of extraneous environmental and design factors invariant with wheel procession.

In a state of complete efficiency the Fill Profile, $A_{load}$, would be conserved throughout the arc of travel and we could express that as:-

$$I_F = \pi A_{load}$$

**Equation B58**

where $I_F$ is the Full Retention Integral.

In our Denny-Warren Model, the perfect state lasts only until $+\pi/2$ radians, at some time after spill begins and is complete substantially before BC eventuates at $\pi$ radians.

This actual profile history is described by:-

$$
\begin{aligned}
I_A &= \frac{\pi}{2}.A_{load} + \sum A.\Delta\psi \\
&= \frac{\pi}{2}.A_{load} + \sum A.(\Delta\theta + \Delta\chi) \\
&= \frac{\pi}{2}.A_{load} + \sum (A_{i-1} - \Delta A_U)(\Delta\theta + \Delta\chi)
\end{aligned}
$$

**Equation B59**

where $A_U$ is the Unwetted Area. Note that in this context $\Delta\theta$ is the finite difference between present and precursor $\theta$s, $\theta_i$ and $\theta_{i-1}$: Not the Denny $\Delta\theta$, Interbucket Angle.

Accordingly:-

$$REF = \frac{I_A}{I_F} = \frac{\frac{\pi}{2}.A_{load} + \sum A.\Delta\psi}{\pi.A_{load}}$$

$$= \frac{1}{2} + \frac{\sum A.\Delta\psi}{\pi.A_{load}}$$

**Equation B60**

REF is the ( dimensionless ) Retentional Efficiency Fraction and is the same thing as the hydraulic efficiency of the bucket waterwheel.

Exhaustion

The Angle of Exhaustion, $\psi_x$, is the angle at which the last drop of water slides over the float board to the tailrace.
It is given by:-

$$\psi_x = \frac{\pi}{2} + \chi_{final} + \theta_{final}$$

$$= \frac{\pi}{2} + \arccos\left[\frac{w_{final}^2 - (R^2 + r^2)}{-2Rr}\right] + \theta_{max}$$

$$= \frac{\pi}{2} + \arccos\left[\frac{f^2 - (R^2 + r^2)}{-2Rr}\right] + \theta_{max}$$

**Equation B61**

$\psi_x$ is therefore a function of design attributes only.
The Empty Angle, $\psi_o$, of redundant wheel procession is therefore:-

$$\psi_o = \pi - \psi_x$$

**Equation B62**

The Character of Water Area Profile Progress

In our Denny-Warren Model of overshot wheel progress we assume that no spill occurs before the bucket progresses from TC to $+\pi/2$ radians ( i.e. the horizontal ), whereupon the Angle of Incipient Spill $\theta_{spill}$ is determined by the relations between the fill profile and the accommodation capacity profile of the container.
This and other design assumptions restrict the unwasteful levels of initial fill depth to a narrow range that varies with the wheel design and the operating conditions of wheel and water.

Known Error Conditions

1       OUTSET OVERFLOW
The attempted Fill Depth exceeds the difference between the Outer and Inner Wheel Radii, R-r , taken to be tantamount to the shroud depth.
Some or most water cascades uselessly over and across the structure.

2       ACCOMMODATION EXCEEDED
The Fill Profile Area is bigger than the designed Accommodation Area.
The water fill is so excessive that some of the water spills from the bucket before $+\pi/2$ is reached. This too wastes potential operating fluid.

3       SKIPAHEAD
Bucket Velocity, $\omega R_E$, exceeds Tangential Water Input Velocity, v.

This seemingly irrational theoretical problem was a source of great confusion to the scientists and theoreticians of the eighteenth and nineteenth centuries. A natural corollary of Borda Theory it seems to arise mathematically from the treatment of torque as a function of weight ( a force ) rather than mass, and under certain flow and wheel speed states can imply a negative torque due to inlet flow impulse. On the face of it, such a negative torque has a diminutive effect on power output. Theoretical skipahead is eliminated if the velocity difference is squared ( i.e. the water impulse is treated as energy ), but this device introduces other problems that also compromise analysis as a reflection of reality. In particular, it was early found that *in a confined channel* the velocity differential varied as the first power and not the second as a dogmatic application of Galilean mechanics might have suggested. Further difficulties arose from the hydrodynamic phenomenon of *vena contracta* in which the spouting nappe of a fluid jet falling through space contracts in sectional area by up to nearly half: Accordingly water impact velocity might be almost double its calculated aperture speed on leaving the launder.

The wheel "skips ahead" of its inlet water nappe and tends to overtake its own power source. In practical water mill operation this is probably impossible because of the existence of an applied payload.

Another confusion involving torque and power is sometimes encountered in literature. Some theoreticians fail to realise that the moment arm operative on an overshot waterwheel is not wheel radius but rather wheel diameter. This leads them to underestimate power output by a factor of two.

Several other problems with waterwheel modelling are known to exist, including modification of the Spill Angle and other water loss determinants by centrifugal and other dynamic effects. Denny adverts to some of these issues, but they are neglected by the present model.

Study of an Ideal Wheel

Several ideal wheels were of course modelled but the present one was devised to represent a typical 24-bucket English overshot plank wheel with a soal radius of 2 meters ( in other words six feet or a fathom, give or take ). It was awarded a very generous bucket depth of nearly 24% of the soal radius, and a rising board rose to 40% of the bucket depth.

Otherwise the design was not too different to that of Splashy Mill in the Moddeshall System.

All dimensions were of course metricated. Table B1 defines the precise dimensions.

**Basic Design Constants**

| | | | | |
|---|---|---|---|---|
| Outer Wheel Radius | $R$ | 2.4761905 | | |
| Inner Wheel Radius | $r$ | 2 | | |
| Bucket Width | $W$ | 1 | | |
| Rising Board ( RB ) Height | $a$ | 0.1904762 | | |
| Bucket Count | $n$ | 24 | | |
| Interbucket Angle | $\Delta\theta$ | 0.2617994 radians | 15.00 degrees | |
| Interboard Angle | $\phi$ | 1.8751923 | 107.44 | |
| | | | | |
| AS/AT Areas Ratio | $A_{rat}$ | 4.2 | | |
| Interbucket Sector Triangle Area | $AS$ | 0.5176381 | | |
| Maximum Soal Contact $\theta$ Sweep | $AT_{max}$ | 0.1232472 | | |
| Rising Board Swept Area | $A_{rising}$ | 0.0610367 | | |
| Float Board ( FB ) Length | $f$ | 0.6717676 | | |
| Maximum Surface Trace Length | $W_{rising0}$ | 0.7511684 | | |
| Apical Angle | $\theta_{max}$ | 1.004601 | 1.004601 | 57.56 degrees |
| Unwetted Area $\theta$ at RB Base | $\theta_{rising0}$ | 0.760261 | | 43.56 degrees |

### Table B1
### Design Parameters of an Ideal Wheel for
### Water Load Profile Progress Studies

In our Denny-Warren Model of Water Profile evolution all diminution of bucket content is confined to the $+\pi/2$ to $\pi$ quadrant. By definition, no Fill Depth exceeds the value that provides an $A_{load}$ conserved at $+\pi/2$. Furthermore, Borda Theory insists that all buckets are empty by the time Bottom Center is reached at $\pi$ radians, and our bucket designs and control schemes engineer that also.

Table B2 presents the relevant system outcomes for the ideal wheel of Table B1, whilst Figure B8 plots local bucket water profile A against Progress Lapse Angle, $\psi$.

**Design Corollaries**

| | | | | |
|---|---|---|---|---|
| Depth of Fill | d | 0.308 | | |
| | | | | |
| Outer Wheel Radius | R | 2.47619 | | |
| Inner Wheel Radius | r | 2 | | |
| Bucket Width | W | 1 | | |
| Rising Board ( RB ) Height | a | 0.190476 | | |
| Bucket Count | n | 24 | | |
| Interbucket Angle | $\Delta\theta$ | 0.261799 | | |
| Interboard Angle | $\phi$ | 1.875192 | | |
| Float Board ( FB ) Length | f | 0.671768 | | |
| | | | | |
| Float Board Inclination | $\alpha$ | 0.304396 | radians | 17.44 degrees |
| Board Displacement Angle | $\zeta$ | 0.566195 | | 32.44 |
| | | | | |
| Radial Depth Function 1 | $\Phi(R,r,d,d,d)$ | 0.388184 | | |
| Radial Depth Function 2 | $\Phi(R,r,a,d,a)$ | 0.368417 | | |
| | | | | |
| Auxiliary Combination Constant | $\Xi_1$ | 0.098598 | | |
| Auxiliary Combination Constant | $\Xi_2$ | 1.035276 | | |
| Auxiliary Combination Constant | $\Xi_3$ | 0.184284 | | |
| | | | | |
| Shallow Load Critical Depth | $d_{crit}$ | 0.115838 | | |
| Left-of-Center Load Polygon | $A_{loadl}$ | 0.162031 | | |
| Right-of-Center Load Polygon | $A_{loadr}$ | 0.021982 | | |
| Shallow Load Polygon | $A_{loads}$ | 0.196026 | | |
| | | | | |
| Fill Profile Area | $A_{load}$ | 0.178053 | | |
| Accommodation Area | $A_{ac}$ | 0.178323 | | |
| Segment Area | $A_{seg}$ | 0.005961 | | |
| Defect Area | $A_{def}$ | 0.00027 | | |
| Spill Angle | $\theta_{spill}$ | 0.002384 | | |
| | | | | |
| Angle of Exhaustion | $\psi_x$ | 2.788721 | | |
| Empty Angle | $\psi_o$ | 0.352872 | | |
| Actual Retention Integral | $I_A$ | 0.414023 | | |
| Full Retention Integral | $I_F$ | 0.55937 | | |
| Retentional Efficiency Fraction | REF | 0.740161 | | |

**Table B2**
**Geometrical Design Outcomes of an Ideal Wheel for**
**Water Load Profile Progress Studies**

Our overall wheel management policy is to keep hydraulic efficiency as high as possible by delaying spill to the maximum practicable $\psi$, and then, at minimum energetic expense, impeding loss as much as is achievable whilst ensuring that all water has left the buckets at Bottom Center.

Broadly, these objectives may be achieved by making the Fill Depth as shallow as manageable, and as is consistent with the evasion of error conditions.

This implies that for maximal power output the wheel should turn briskly. But not so fast that centrifugal forces cause premature water expulsion or "skipahead" issues, however those may manifest.

Contrary to what is often stated, angular speed should be as high as optimal for the load, not as low as practicable, and of course modern turbines are designed to turn at an optimal high speed with minimum torque, as is indeed best for the electrical equipment to which they are coupled.

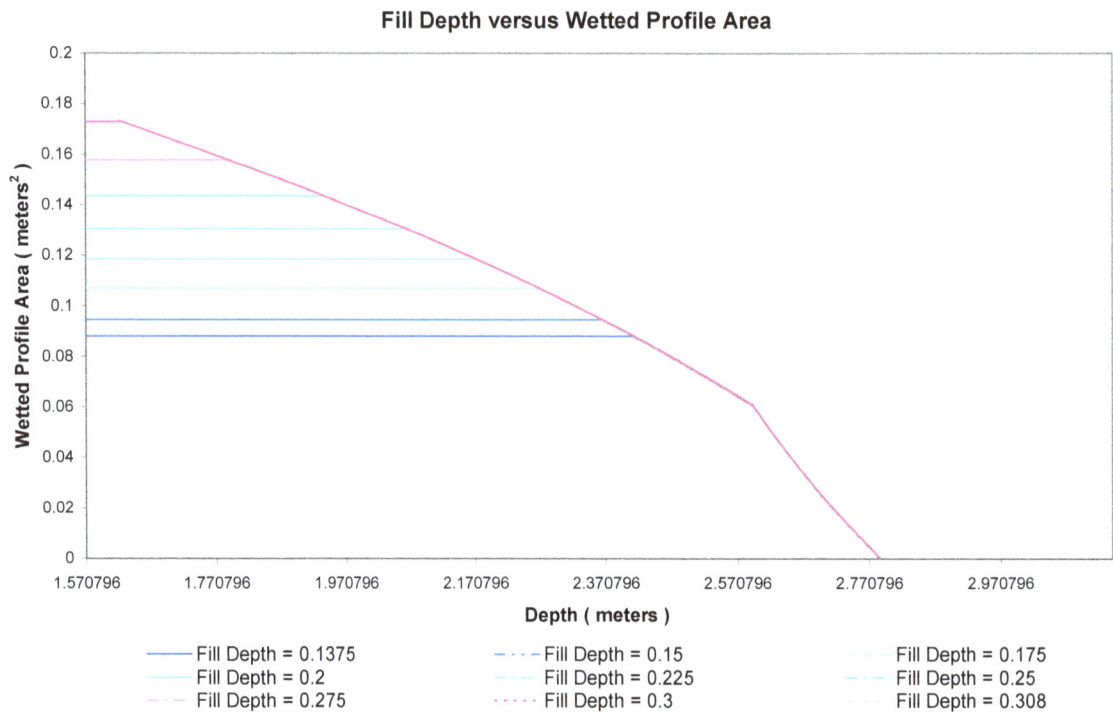

**Fill Depth versus Wetted Profile Area**

Legend:
- Fill Depth = 0.1375
- Fill Depth = 0.15
- Fill Depth = 0.175
- Fill Depth = 0.2
- Fill Depth = 0.225
- Fill Depth = 0.25
- Fill Depth = 0.275
- Fill Depth = 0.3
- Fill Depth = 0.308

**Figure B8**
**The Conservation and Decline of**
**Bucket Water Load Profile with Progress Angle**
**For Bucket Emptying after Various TC Fill Depths**

It is clear from Figure B8 that bucket water profile area, and by implication bucket load is longer conserved for shallower original fill depths, but that when the relevant $\theta_{spill}$ is eventually attained, the process of bucket emptying is wholly determined by bucket design geometry.

## Fill Depth – Geometry Relations

As has been seen elsewhere, wheel bucket geometry exhibits some remarkably simple relationships modelled to a high degree of approximation by elementary algebraic polynomials.

Further research is needed to establish the scientific and design bases of these relationships.

### Depth and Retentional Efficiency Fraction

This key relationship for wheel management practice can be fitted with a simple cubic regression:-

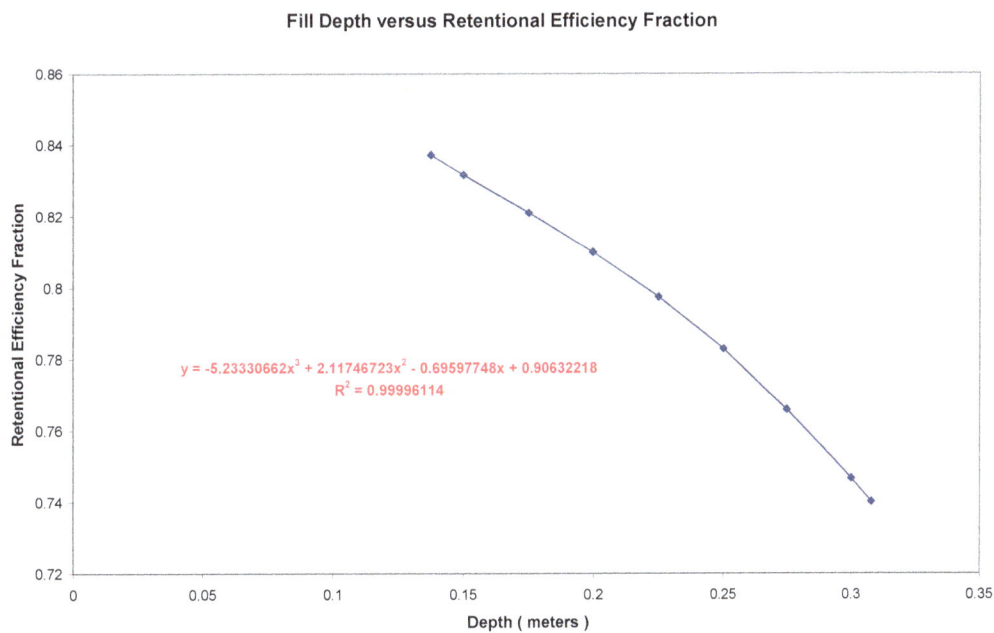

**Fill Depth versus Retentional Efficiency Fraction**

$y = -5.23330662x^3 + 2.11746723x^2 - 0.69597748x + 0.90632218$
$R^2 = 0.99996114$

**Figure B9**
**Depth and Retentional Efficiency Fraction**

### Depth and Bucket Fill Volume

This too is extremely closely matched by a cubic equation, but may even be modelled by a linear fitment to the limits of operational error.

**Fill Depth versus Bucket Fill Volume**

$$y = 4.26559387x^3 - 2.36912877x^2 + 0.91707383x - 0.00414203$$
$$R^2 = 0.99996253$$

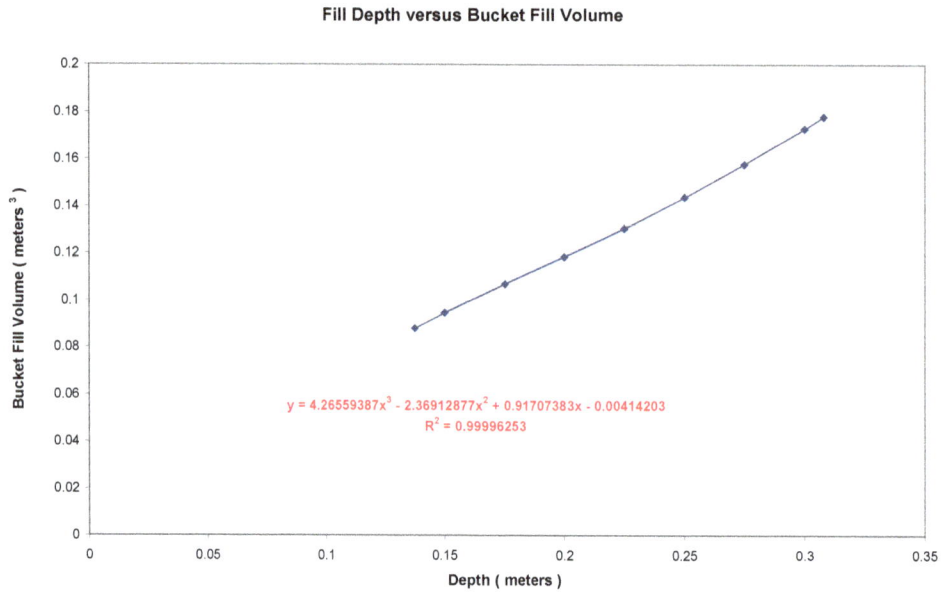

**Figure B10**
**Depth and Bucket Fill Volume**

Depth and Angular Speed

This cubic shows how, for a given Flow Rate Q, angular speed ω in radians per second reduces with increasing fill depth.

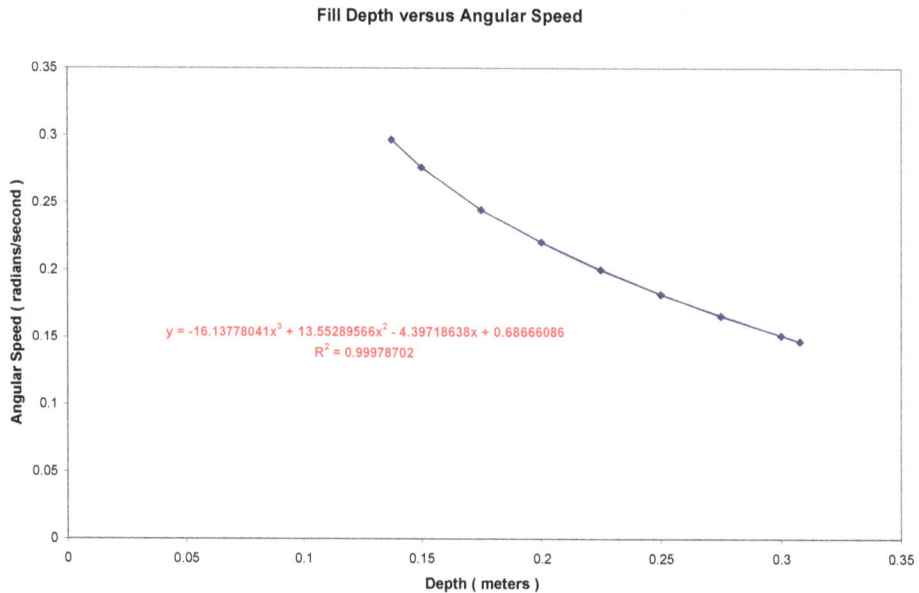

**Fill Depth versus Angular Speed**

$$y = -16.13778041x^3 + 13.55289566x^2 - 4.39718638x + 0.68666086$$
$$R^2 = 0.99978702$$

**Figure B11**
**Depth and Angular Speed**

Depth and Kinetic/Gravitational Power Ratio

This plot illustrates the variation, for equivalent Flow Rate, of The Ratio of Impulse to Gravitational Power Output $P_w/P_g$.

It is highly counterintuitive in that the importance of kinetic energy input is greater for the increased fill depths.

This apparent anomaly is worthy of further analysis.

**Fill Depth versus Kinetic/Gravitational Power Ratio $P_w/P_g$**

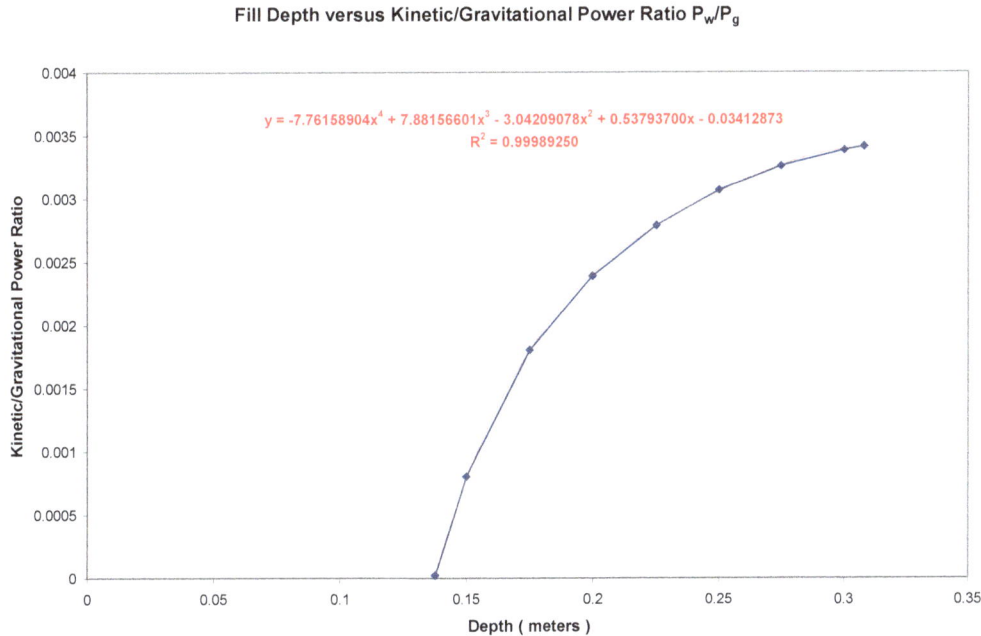

$$y = -7.76158904x^4 + 7.88156601x^3 - 3.04209078x^2 + 0.53793700x - 0.03412873$$
$$R^2 = 0.99989250$$

**Figure B12**
**Depth and Power Ratio**

Depth and Spill Angle

A simple quadratic seems adequate to model the dramatic decline of Incipient Spill Angle $\theta_{spill}$ with Fill Depth. To an extent this relation harmonises with the Depth-REF relations of Figure B9.

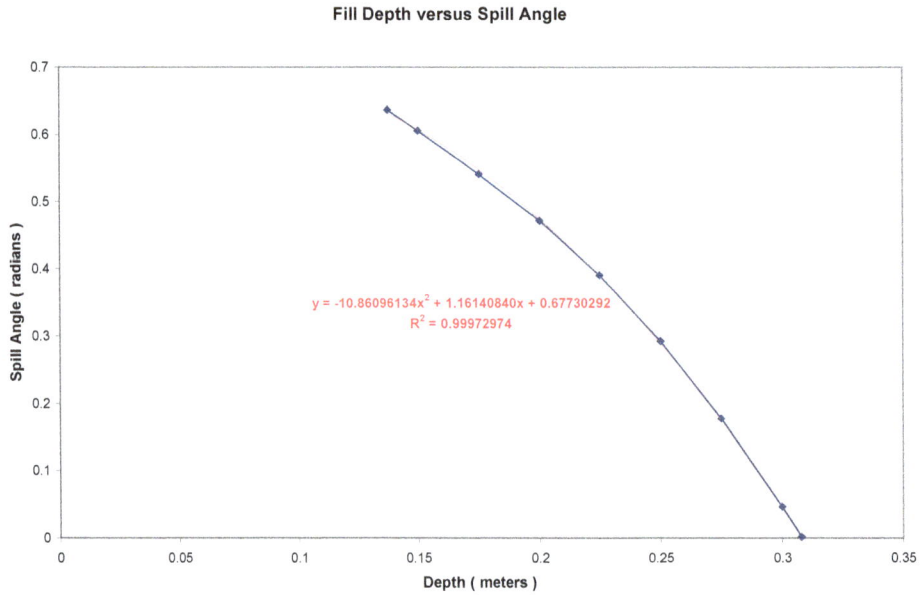

$$y = -10.86096134x^2 + 1.16140840x + 0.67730292$$
$$R^2 = 0.99972974$$

**Figure B13: Depth and Spill Angle**

Depth versus Influx and Bucket Velocities

For a given Flow Rate and gate configuration the Influx Velocity is of course the constant illustrated by the upper horizontal line.

A cubic regression equation models the reduction of the bucket speed, and the developing differential may explain some or all of the paradox that kinetic energy input increases with fill depth ( because at constant Q a deeper fill implies a slower bucket ).

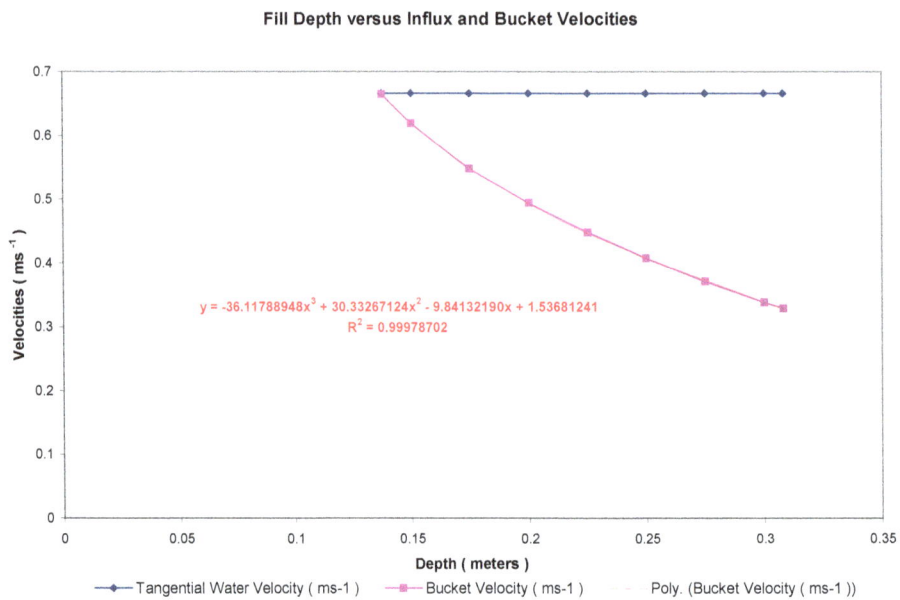

$$y = -36.11788948x^3 + 30.33267124x^2 - 9.84132190x + 1.53681241$$
$$R^2 = 0.99978702$$

Tangential Water Velocity ( ms-1 )     Bucket Velocity ( ms-1 )     Poly. (Bucket Velocity ( ms-1 ))

**Figure B14**
**Depth versus Influx and Bucket Velocities**

Depth and Power Output Metrics

This chart shows the remarkable and counterintuitive decline of overshot waterwheel power output with increasing bucket water fill depth, giving that Flow Rate is constant.

It is clear that all the candidate curves are capable of substantially cubic regression modelling, and in the case of the Denny-Warren Model power profile the fitted cubic equation has an R=0.99996982, indicating that the fitted cubic accounts for 99.9939641 percent of the variation in the data. We must assume that a cubic equation is a deterministic descriptor of the Depth-Power linkage and that analysis will disclose its terms.

Note that the Borda and Denny-Warren power models are probably equivalent. All honor to the chevalier who developed his waterwheel model without the benefit of computer assistance!

Interestingly, the Franklin Institute model reports substantially smaller power estimates, raising the possibility of assessments under standard loads?

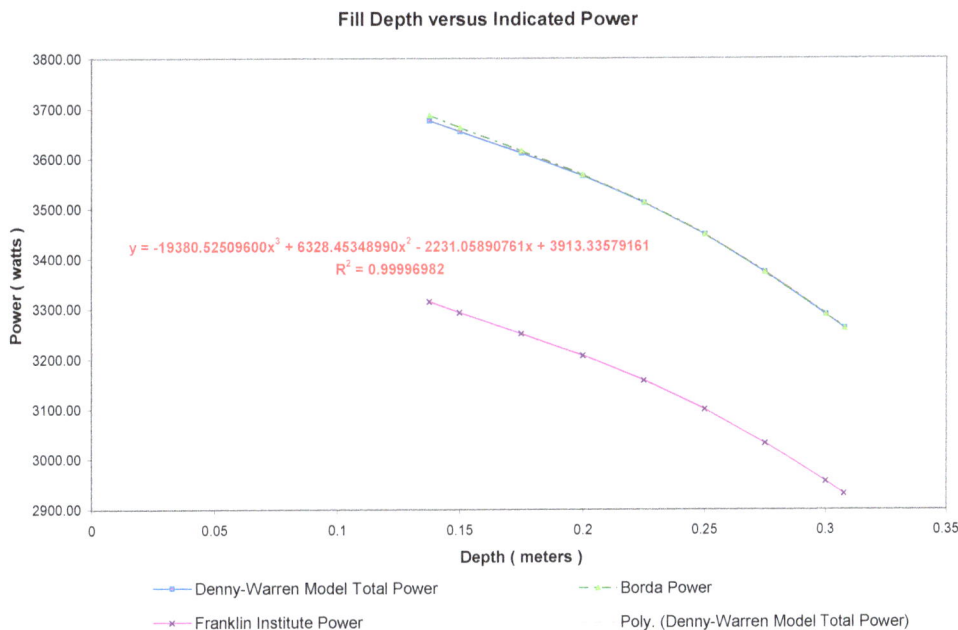

Fill Depth versus Indicated Power

$$y = -19380.52509600x^3 + 6328.45348990x^2 - 2231.05890761x + 3913.33579161$$
$$R^2 = 0.99996982$$

**Figure B15**
**Depth and Power Output Metrics**

## The Calculation of Spill Angles

We have seen that the general Wheel Progress Angle, $\psi$, is intimately related to the Bucket Water Surface Angle, $\theta$, and the Gyratory Co-Angle, $\chi$.

The values of these angles at incipient bucket spill can in principle be calculated using the same formulae.

Let us define:-

$\chi_{spill}$ = Gyratory Co-Angle of Incipient Spill

$\theta_{spill}$ = Bucket Water Surface Spill Angle
( Critical Angle of Incipient Spill )

$\psi_{spill}$ = Wheel Progress Lapse Angle of Incipient Spill

$w_{spill}$ = Length of Bucket Water Surface Profile at
Incipient Spill

As before, R is the Wheel Outer ( Shroud ) Radius, r is the Wheel Inner ( Soal ) Radius; $A_{def}$ is the Defect Profile Area; $A_{load}$ is the Wetted Profile at Fill and $A_{ac}$ is the designed Bucket Accommodation Profile Area.

$A_{seg}$ is the corresponding Soal Segment Area and although it does not directly affect the present computations, it is possibly advisable to discount it from $A_{def}$ at the outset, in more accurate practical estimates of the spill angles.

Firstly, $\theta_{spill}$ is given by:-

$$\theta_{spill} = \arcsin\left[\frac{\dfrac{2.A_{def}}{(R-r)}}{\sqrt{\left(R^2+r^2\right)-2Rr\left[\dfrac{\sqrt{r^2(R-r)^2-4A_{def}^2}}{r(R-r)}\right]}}\right]$$

**Equation B63**

Therefore:-

$$\theta_{spill} = \arcsin\left[\dfrac{\dfrac{2.\left(A_{ac} - A_{load}\right)}{\left(R - r\right)}}{\sqrt{\left(R^2 + r^2\right) - 2Rr\left[\dfrac{\sqrt{r^2\left(R - r\right)^2 - 4A_{def}^{\,2}}}{r\left(R - r\right)}\right]}}\right]$$

$$= \arcsin\left[\dfrac{2.\left(A_{ac} - A_{load}\right)}{\left(R - r\right)\sqrt{\left(R^2 + r^2\right) - 2Rr\left[\dfrac{\sqrt{r^2\left(R - r\right)^2 - 4A_{def}^{\,2}}}{r\left(R - r\right)}\right]}}\right]$$

**Equation B64**

Noting that:-

$$w_{spill} = \dfrac{2.A_{def}}{\left(R - r\right).\sin\left(\theta_{spill}\right)}$$

**Equation B65**

we may define $\chi_{spill}$ using:-

$$\chi_{spill} = \arccos\left[\dfrac{w_{spill}^{\,2} - \left(R^2 + r^2\right)}{-2Rr}\right]$$

$$= \arccos\left[\dfrac{\left[\dfrac{2.A_{def}}{\left(R - r\right)\sin\left(\theta_{spill}\right)}\right]^2 - \left(R^2 + r^2\right)}{-2Rr}\right]$$

**Equation B66**

To obtain the Progress Angle at Spill we can now use:-

$$\psi_{spill} = \dfrac{\pi}{2} + \theta_{spill} + \chi_{spill}$$

**Equation B67**

An alternative and more economical route to $\chi_{spill}$ is available based upon our knowledge that the ratio of the triangles AS and AT, Arat, is equivalent to r/(R-r); and also the Sine Rule:-

$$AS = \frac{1}{2} r.r.\sin\left(\chi_{spill}\right)$$

**Equation B68**

$$AT = \frac{1}{2} w.(R-r).\sin\left(\theta_{spill}\right)$$

**Equation B69**

Therefore:-

$$Arat = \frac{AS}{AT} = \frac{\frac{1}{2} r.r.\sin\left(\chi_{spill}\right)}{\frac{1}{2} w.(R-r).\sin\left(\theta_{spill}\right)} = \frac{r.\sin\left(\chi_{spill}\right)}{w.\sin\left(\theta_{spill}\right)} \cdot \frac{r}{(R-r)} = \eta \cdot \frac{r}{(R-r)} = 1 \cdot \frac{r}{(R-r)}$$

**Equation B70**

Because the ratio $\eta$ is unity we may rearrange its variables to yield:-

$$\chi_{spill} = \arcsin\left[\frac{w.\sin\left(\theta_{spill}\right)}{r}\right]$$

**Equation B71**

An Approximation for Retentional Efficiency Fraction

Analysis of the Profile Progress Diagram of local wetted area against progress angle enables us to formulate the following rough estimator of the Retentional Efficiency Fraction, REF:-

$$REF_{approx} = \frac{\psi_{spill}.A_{load} + \frac{1}{2}\left(\psi_x - \psi_{spill}\right)}{\pi.A_{load}} = \frac{\psi_{spill} - \psi_x}{2\pi}$$

**Equation B72**

where $\psi_x$ is the Exhaustion Angle.

$REF_{approx}$ will show a negligible error for the shallower legitimate depths up to about seven percent for the largest acceptable fill depths.

Figure B16 shows the REF and $REF_{approx}$ curves versus Fill Depth for the data of Table B1.

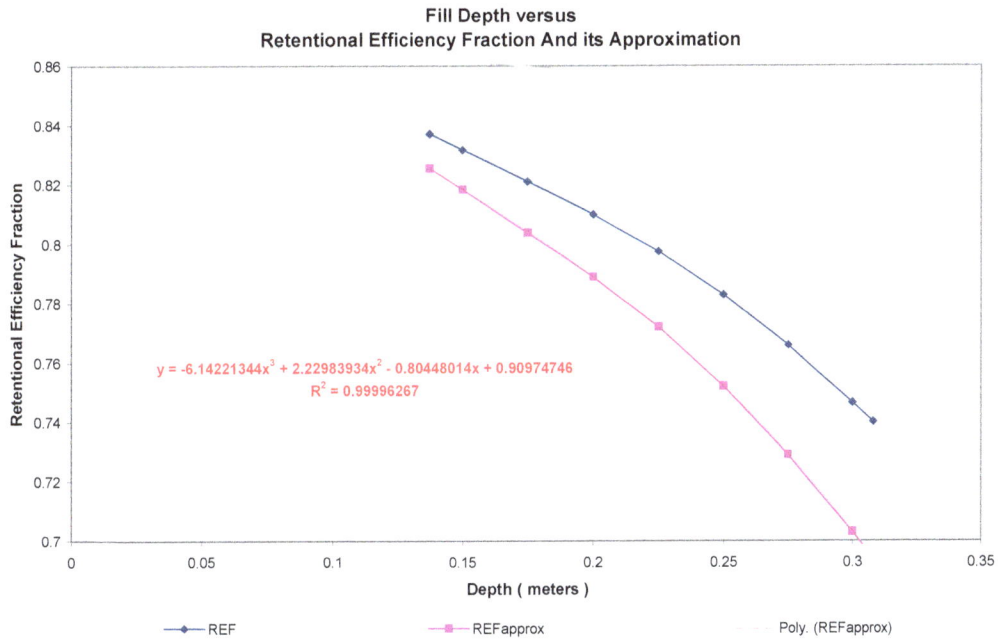

**Fill Depth versus
Retentional Efficiency Fraction And its Approximation**

$y = -6.14221344x^3 + 2.22983934x^2 - 0.80448014x + 0.90974746$
$R^2 = 0.99996267$

REF — REFapprox — Poly. (REFapprox)

**Figure B16
The Variation of REF and REF$_{approx}$ with Fill Depth**

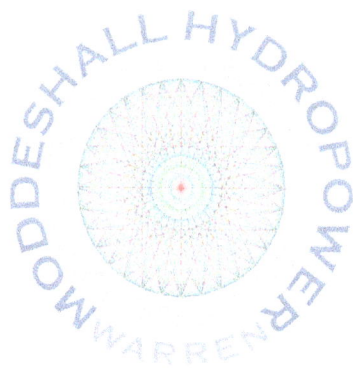

**Dimensions, Histories and Uses**
**of the**
**Moddeshall Valley Mills**

| Mill Name | Grid Reference | Latitude ( MultiMap ) | | | Longitude ( Multimap ) | | | Decimal Latitude | Decimal Longitude | Latitude ( GoogleEarth ) | | | | Longitude ( GoogleEarth ) | | | |
|---|---|---|---|---|---|---|---|---|---|---|---|---|---|---|---|---|---|
| | | ° | ' | " | ° | ' | " | | | ° | ' | " | "/100 | ° | ' | " | "/100 |
| Boar | SJ926368 | 52 | 55 | 43 | 2 | 6 | 39 | 52.92857 | -2.11092 | 52 | 55 | 44 | 16 | 2 | 7 | 41 | 98 |
| Lower Moddershall | SJ921366 | 52 | 55 | 37 | 2 | 7 | 8 | 52.92694 | -2.11882 | 52 | 55 | 37 | 11 | 2 | 7 | 7 | 19 |
| Splashy | SJ919366 | 52 | 55 | 37 | 2 | 7 | 14 | 52.9269 | -2.1206 | 52 | 55 | 37 | 3 | 2 | 7 | 14 | 20 |
| Ochre | SJ918365 | 52 | 55 | 36 | 2 | 7 | 19 | 52.92672 | -2.12201 | 52 | 55 | 35 | 12 | 2 | 7 | 20 | 66 |
| Mosty Lee | SJ918362 | 52 | 55 | 24 | 2 | 7 | 26 | 52.92342 | -2.12386 | 52 | 55 | 24 | 65 | 2 | 7 | 25 | 77 |
| Wetmore | SJ916360 | 52 | 55 | 15 | 2 | 7 | 33 | 52.92088 | -2.12573 | 52 | 55 | 15 | 88 | 2 | 7 | 33 | 6 |
| Vanity | SJ916357 | 52 | 55 | 7 | 2 | 7 | 32 | 52.91849 | -2.12558 | 52 | 55 | 6 | 9 | 2 | 7 | 32 | 33 |
| Ivy | SJ916355 | 52 | 54 | 54 | 2 | 7 | 32 | 52.915 | -2.12559 | 52 | 54 | 58 | 99 | 2 | 7 | 34 | 98 |
| Hayes | SJ912351 | 52 | 54 | 48 | 2 | 7 | 55 | 52.9132 | -2.13191 | 52 | 54 | 46 | 41 | 2 | 7 | 55 | 89 |
| Coppice | SJ908347 | 52 | 54 | 35 | 2 | 8 | 16 | 52.90967 | -2.13772 | 52 | 54 | 34 | 53 | 2 | 8 | 16 | 50 |
| Weaver | SJ906339 | 52 | 54 | 11 | 2 | 8 | 25 | 52.90308 | -2.14041 | 52 | 54 | 12 | 60 | 2 | 8 | 16 | 50 |
| Stone | SJ905340 | 52 | 52 | 12 | 2 | 8 | 32 | 52.90341 | -2.14227 | 52 | 54 | 12 | 56 | 2 | 8 | 32 | 23 |

| Mill Name | Grid Reference | Elevation (feet) (Virtual Earth) | | | Elevation (meters) (Virtual Earth) | | | Elevation (meters) (GoogleEarth) | | |
|---|---|---|---|---|---|---|---|---|---|---|
| | | Head | Mill | Tail | Head | Mill | Tail | Head | Mill | Tail |
| Boar | SJ926368 | 518 | 528 | 512 | 157.9 | 160.9 | 156.1 | 169 | 166 | 163 |
| Lower Moddershall | SJ921366 | | 459 | | 0.0 | 139.9 | 0.0 | | 145 | |
| Splashy | SJ919366 | 453 | 456 | 443 | 138.1 | 139.0 | 135.0 | 143 | 140 | 137 |
| Ochre | SJ918365 | 441 | 429 | 429 | 134.4 | 130.8 | 130.8 | 137 | 138 | 138 |
| Mosty Lee | SJ918362 | 419 | 422 | 421 | 127.7 | 128.6 | 128.3 | 136 | 136 | 136 |
| Wetmore | SJ916360 | 444 | 417 | 425 | 135.3 | 127.1 | 129.5 | 140 | 139 | 140 |
| Vanity | SJ916357 | | 420 | | 0.0 | 128.0 | 0.0 | | 134 | |
| Ivy | SJ916355 | 417 | 436 | 394 | 127.1 | 132.9 | 120.1 | 129 | 129 | 129 |
| Hayes | SJ912351 | 358 | 356 | 350 | 109.1 | 108.5 | 106.7 | 113 | 113 | 112 |
| Coppice | SJ908347 | 359 | 356 | 351 | 109.4 | 108.5 | 107.0 | 116 | 115 | 114 |
| Weaver | SJ906339 | 307 | 298 | 296 | 93.6 | 90.8 | 90.2 | 99 | 98 | 97 |
| Stone | SJ905340 | 347 | 298 | 296 | 105.8 | 90.8 | 90.2 | 98 | 105 | 97 |

| Mill Name | Grid Reference | River | Year of Build | Year of Closure | Design Use | Use in c1860 | Final Use |
|---|---|---|---|---|---|---|---|
| Boar | SJ926368 | Moddershall Brook | 1798 | 1954 | corn | flint | flint |
| Lower Moddershall | SJ921366 | Moddershall Brook | | | corn | flint | |
| Splashy | SJ919366 | Moddershall Brook | 1752 | 1958 | corn | flint | bone |
| Ochre | SJ918365 | Moddershall Brook/Scotch Brook | 1829 | 1946 | flint | flint and color | flint |
| Mosty Lee | SJ918362 | Scotch Brook | 1716 | 1961 | corn | flint | flint and color |
| Wetmore | SJ916360 | Scotch Brook | 1763 | 1960 | oil and flint | flint and oil | bone |
| Vanity | SJ916357 | Scotch Brook | 1775 | 1818 | | | |
| Ivy | SJ916355 | Scotch Brook | 1740 | 1966 | oil | flint | bone |
| Hayes | SJ912351 | Scotch Brook | 1750 | 1966 | flint | flint | |
| Coppice | SJ908347 | Scotch Brook | 1720 | 1953 | paper | flint | flint |
| Weaver | SJ906339 | Scotch Brook | 1775 | 1976 | corn | corn and oats | corn and oats |
| Stone | SJ905340 | Scotch Brook | | | | | |

| Mill Name | Grid Reference | Wheel Type | Data from Sketches (feet) | | Data from Text (feet) | | Data from Photos (mm) | Radius | Bucket Annulus (feet) | Number of Buckets |
|---|---|---|---|---|---|---|---|---|---|---|
| | | | Diameter | Width | Diameter | Width | Annulus Depth | | | |
| Boar | SJ926368 | High Breast | 20 | 4.5 | 20 | 4.5 | | | | |
| Lower Moddershall | SJ921366 | Overshot | 9 | | | | | | | |
| Splashy | SJ919366 | Overshot | 16 | 5.5 | 16 | 6 | 6.2 | 50 | 0.992 | 24 |
| Ochre | SJ918365 | Overshot | 12 | 9 | 12 | 12 | | | | |
| Mosty Lee | SJ918362 | Pitchback | 18 | 6.5 | 18 | 6.5 | 9 | 150 | 0.54 | 56 |
| Wetmore | SJ916360 | High Breast | 21 | 7 | 21 | 7 | | | | |
| Vanity | SJ916357 | | | | | | | | | |
| Ivy | SJ916355 | High Breast | 19 | 6 | 19 | 6 | 16.8 | 107 | 1.491589 | 84 |
| Hayes | SJ912351 | Overshot | 20 | 6.25 | 20 | 5.1 | | | | |
| Coppice | SJ908347 | High Breast | 20 | 6.5 | 20 | 6.5 | 4 | 36.5 | 1.09589 | 108 |
| Weaver | SJ906339 | Overshot | | | 24 | 5.75 | | | | |
| Stone | SJ905340 | | | | | | | | | |

| Mill Name | Grid Reference | Wheel Type | Notes |
|---|---|---|---|
| Boar | SJ926368 | High Breast | |
| Lower Moddershall | SJ921366 | Overshot | |
| Splashy | SJ919366 | Overshot | originally 2*pitchback 18' dia |
| Ochre | SJ918365 | Overshot | |
| Mosty Lee | SJ918362 | Pitchback | |
| Wetmore | SJ916360 | High Breast | |
| Vanity | SJ916357 | | |
| Ivy | SJ916355 | High Breast | |
| Hayes | SJ912351 | Overshot | |
| Coppice | SJ908347 | High Breast | |
| Weaver | SJ906339 | Overshot | second wheel 22*6ft for nocturnal flint grinding |
| Stone | SJ905340 | | |

# DATA APPENDIX D

## Wheel Performance Results
## of the
## Moddeshall Valley Mills

| Name | | Boar Mill | Splashy Mill | Ochre Mill |
|---|---|---|---|---|
| R | | 3.048 | 2.4384 | 1.8288 |
| r | | 2.744906646 | 2.1360384 | 1.646943988 |
| W | | 1.3716 | 1.6764 | 2.7432 |
| a | | 0 | 0 | 0 |
| n | | 24 | 24 | 48 |
| | | | | |
| Cumulative Q | | 0.049288437 | 0.056656256 | 0.099186518 |
| Nom. Stream Reach Power | | 19525.20022 | 14946.05417 | 14946.05417 |

**Influx Parameters**

| | | | | |
|---|---|---|---|---|
| Depth of Fill | d | 0.0335 | 0.02595 | 0.0205 |
| Approximate Half-Full Depth | | 0.151546677 | 0.1511808 | 0.090928006 |
| Water Influx ( $m^3s^{-1}$ ) | $Q_{in}$ | 0.049288437 | 0.056656256 | 0.099186518 |
| Sluice Gate Area ( $m^2$ ) | $A_G$ | 0.06 | 0.06 | 0.06 |

**Consequential Parameters**

| | | | | |
|---|---|---|---|---|
| Tangential Water Velocity ( $ms^{-1}$ ) | v | 0.821473946 | 0.944270929 | 1.653108625 |
| Bucket Velocity ( $ms^{-1}$ ) | $\omega R_E$ | 0.811828401 | 0.943343093 | 1.625478116 |
| Indicated Velocity Head | h | 0.034383009 | 0.04543071 | 0.139238415 |
| Effective Wheel Radius | $R_E$ | 2.896453323 | 2.2872192 | 1.737871994 |
| Fill Profile Area | $A_{load}$ | 0.033565142 | 0.021452453 | 0.005060232 |
| Bucket Fill Volume | $V_f$ | 0.046037949 | 0.035962892 | 0.013881229 |
| Bucket Presentation Duration ( s ) | $t_f$ | 0.93405171 | 0.634755892 | 0.139950769 |
| Cycle Time ( s ) | T | 22.41724105 | 15.23414141 | 6.717636902 |
| Bucket Fall Time  ( Half-Cycle ) | $T_F$ | 11.20862052 | 7.617070705 | 3.358818451 |
| Angular Velocity ( cycles/second ) | f | 0.044608522 | 0.065642032 | 0.148861871 |
| Angular Speed ( radians/s ) | $\omega$ | 0.280283613 | 0.412441052 | 0.935326723 |
| Fill Weight per Second | K | 483.5087548 | 555.7854429 | 972.9979836 |
| Effective Fall | H-h | 5.116201529 | 3.94124074 | 2.709462708 |

**Efficiency Metrics**

| | | | | |
|---|---|---|---|---|
| Denny Spill Efficiency | $\varepsilon_D$ | 0.984778475 | 0.972303969 | 0.91032861 |
| Denny Combined Energy Efficiency | $\varepsilon_C$ | 0.375339261 | 0.332596382 | 0.206035801 |
| Retentional Efficiency Fraction | REF | 0.883183839 | 0.861579148 | 0.779534603 |
| Approximate REF | $REF_{approx}$ | 0.876748959 | 0.856460255 | 0.776743509 |

**Hydraulic Power Outputs**

| | | | | |
|---|---|---|---|---|
| Torque due to Bucket Water Gravitation | $\tau_g$ | 4412.902001 | 2655.511886 | 1409.294575 |
| Torque due to Water Entry Impulse | $\tau_w$ | 1.376524404 | 0.120191147 | 4.761073609 |
| Power due to Bucket Water Gravitation | $P_g$ | 2473.72823 | 2190.484231 | 2636.301752 |
| Power due to Water Entry Impulse | $P_w$ | 0.385817233 | 0.049571763 | 4.453159375 |
| Denny Gravitational Power In | $P_{gD}$ | 2800.921079 | 2542.406272 | 3381.891892 |
| Denny Kinetic Power In | $P_{wD}$ | 16.62448564 | 25.24972737 | 135.4786971 |
| | | | | |
| Denny Simplified Power Input | $P_D$ | 2817.545565 | 2567.656 | 3517.370589 |
| Denny-Warren Model Total Power | G | 2474.114048 | 2190.533802 | 2640.754911 |
| Borda Power | $P_B$ | 2482.040473 | 2203.109094 | 2704.0411 |
| Franklin Institute Power | $P_{FI}$ | 2231.010263 | 1978.505731 | 2410.605612 |
| | | | | |
| Gate Power ( watts ) | G | 2474.114048 | 2190.533802 | 2640.754911 |
| | | | | |
| Modelled Power Split ( $P_w/P_g$ ) | | 0.000155966 | 2.26305E-05 | 0.001689169 |

**Regional Constants**

| | | | | |
|---|---|---|---|---|
| Water Density ( $kgm^{-3}$ ) | $\rho$ | 999.6445 | 999.6445 | 999.6445 |
| Acceleration Due to Gravity ( $ms^{-2}$ ) | g | 9.8132693 | 9.8132693 | 9.8132693 |
| Specific Weight of Water | $\gamma$ | 9809.780683 | 9809.780683 | 9809.780683 |

**Geometrical Corollarys**

| | | | | |
|---|---|---|---|---|
| Spill Angle | $\theta_{spill}$ | 0.892048082 | 0.831125063 | 0.682781203 |
| Angle of Exhaustion | $\psi_x$ | 2.894211246 | 2.807194286 | 2.533356549 |

| Name | | Mosty Lee Mill | Wetmore Mill | Ivy Mill |
|---|---|---|---|---|
| R | | 2.7432 | 3.2004 | 2.8956 |
| r | | 2.578608 | 2.882151978 | 2.440963738 |
| W | | 1.9812 | 2.1336 | 1.8288 |
| a | | 0 | 0 | 0 |
| n | | 56 | 60 | 84 |
| | | | | |
| **Cumulative Q** | | 0.119408816 | 0.130481902 | 0.1378789 |
| **Nom. Stream Reach Power** | | 14946.05417 | 14946.05417 | 11068.20193 |

### Influx Parameters

| | | Mosty Lee Mill | Wetmore Mill | Ivy Mill |
|---|---|---|---|---|
| Depth of Fill | $d$ | 0.02875 | 0.025875 | 0.0339 |
| Approximate Half-Full Depth | | 0.082296 | 0.159124011 | 0.227318131 |
| Water Influx ( $m^3s^{-1}$ ) | $Q_{in}$ | 0.119408816 | 0.130481902 | 0.1378789 |
| Sluice Gate Area ( $m^2$ ) | $A_G$ | 0.06 | 0.06 | 0.06 |

### Consequential Parameters

| | | Mosty Lee Mill | Wetmore Mill | Ivy Mill |
|---|---|---|---|---|
| Tangential Water Velocity ( $ms^{-1}$ ) | $v$ | 1.990146926 | 2.174698365 | 2.297981661 |
| Bucket Velocity ( $ms^{-1}$ ) | $\omega R_E$ | 1.976019287 | 2.171662686 | 2.297593189 |
| Indicated Velocity Head | $h$ | 0.201802512 | 0.240965209 | 0.269060165 |
| Effective Wheel Radius | $R_E$ | 2.660904 | 3.041275989 | 2.668281869 |
| Fill Profile Area | $A_{load}$ | 0.009106216 | 0.008968697 | 0.006549238 |
| Bucket Fill Volume | $V_f$ | 0.018041235 | 0.019135612 | 0.011977246 |
| Bucket Presentation Duration ( s ) | $t_f$ | 0.151087964 | 0.146653382 | 0.086867866 |
| Cycle Time ( s ) | $T$ | 8.460925979 | 8.799202902 | 7.296900739 |
| Bucket Fall Time ( Half-Cycle ) | $T_F$ | 4.230462989 | 4.399601451 | 3.648450369 |
| Angular Velocity ( cycles/second ) | $f$ | 0.118190373 | 0.113646658 | 0.137044484 |
| Angular Speed ( radians/s ) | $\omega$ | 0.742612017 | 0.71406301 | 0.861075891 |
| Fill Weight per Second | $K$ | 1171.374292 | 1279.998841 | 1352.561766 |
| Effective Fall | $H-h$ | 4.356582429 | 4.552366866 | 3.378955523 |

### Efficiency Metrics

| | | Mosty Lee Mill | Wetmore Mill | Ivy Mill |
|---|---|---|---|---|
| Denny Spill Efficiency | $\varepsilon_D$ | 0.9508098 | 0.871848028 | 0.718042453 |
| Denny Combined Energy Efficiency | $\varepsilon_C$ | 0.273368934 | 0.159425291 | 0.047645057 |
| Retentional Efficiency Fraction | REF | 0.818628261 | 0.748430409 | 0.633170648 |
| Approximate REF | $REF_{approx}$ | 0.813492755 | 0.747178593 | 0.632847498 |

### Hydraulic Power Outputs

| | | Mosty Lee Mill | Wetmore Mill | Ivy Mill |
|---|---|---|---|---|
| Torque due to Bucket Water Gravitation | $\tau_g$ | 3435.972313 | 4080.189166 | 2653.799796 |
| Torque due to Water Entry Impulse | $\tau_w$ | 4.487255381 | 1.204224812 | 0.142868 |
| Power due to Bucket Water Gravitation | $P_g$ | 5103.18866 | 5827.024312 | 4570.24605 |
| Power due to Water Entry Impulse | $P_w$ | 3.332289769 | 0.859892394 | 0.12302019 |
| Denny Gravitational Power In | $P_{gD}$ | 6233.82908 | 7785.659482 | 7218.032076 |
| Denny Kinetic Power In | $P_{wD}$ | 236.3862745 | 308.4351886 | 363.920492 |
| | | | | |
| Denny Simplified Power Input | $P_D$ | 6470.215355 | 8094.094671 | 7581.952568 |
| Denny-Warren Model Total Power | G | 5106.520949 | 5827.884205 | 4570.36907 |
| Borda Power | $P_B$ | 5221.381797 | 5981.241907 | 4752.206296 |
| Franklin Institute Power | $P_{FI}$ | 4659.057951 | 5330.683734 | 4215.119183 |
| | | | | |
| Gate Power ( watts ) | G | 5106.520949 | 5827.884205 | 4570.36907 |
| | | | | |
| Modelled Power Split ( $P_w/P_g$ ) | | 0.000652982 | 0.00014757 | 2.69176E-05 |

### Regional Constants

| | | Mosty Lee Mill | Wetmore Mill | Ivy Mill |
|---|---|---|---|---|
| Water Density ( $kgm^{-3}$ ) | $\rho$ | 999.6445 | 999.6445 | 999.6445 |
| Acceleration Due to Gravity ( $ms^{-2}$ ) | $g$ | 9.8132693 | 9.8132693 | 9.8132693 |
| Specific Weight of Water | $\gamma$ | 9809.780683 | 9809.780683 | 9809.780683 |

### Geometrical Corollarys

| | | Mosty Lee Mill | Wetmore Mill | Ivy Mill |
|---|---|---|---|---|
| Spill Angle | $\theta_{spill}$ | 0.78081973 | 0.631150042 | 0.320868458 |
| Angle of Exhaustion | $\psi_x$ | 2.694295832 | 2.409378547 | 2.022039831 |

| Name | | Hayes Mill | Coppice Mill | Stone Mill |
|---|---|---|---|---|
| R | | 3.048 | 3.048 | 3.6576 |
| r | | 2.744906646 | 2.713972603 | 3.293887975 |
| W | | 1.905 | 1.9812 | 1.7526 |
| a | | 0 | 0 | 0 |
| n | | 96 | 108 | 128 |
| | | | | |
| Cumulative Q | | 0.143734161 | 0.330451825 | 0.34153125 |
| Nom. Stream Reach Power | | 5409.338021 | 12142.74799 | 26243.92343 |

### Influx Parameters

| | | Hayes Mill | Coppice Mill | Stone Mill |
|---|---|---|---|---|
| Depth of Fill | $d$ | 0.031705 | 0.063 | 0.070825 |
| Approximate Half-Full Depth | | 0.151546677 | 0.167013699 | 0.181856012 |
| Water Influx ( $m^3s^{-1}$ ) | $Q_{In}$ | 0.143734161 | 0.330451825 | 0.34153125 |
| Sluice Gate Area ( $m^2$ ) | $A_G$ | 0.06 | 0.12 | 0.12 |

### Consequential Parameters

| | | Hayes Mill | Coppice Mill | Stone Mill |
|---|---|---|---|---|
| Tangential Water Velocity ( $ms^{-1}$ ) | $v$ | 2.395569344 | 2.753765209 | 2.846093747 |
| Bucket Velocity ( $ms^{-1}$ ) | $\omega R_E$ | 2.395521057 | 2.744038527 | 2.845607927 |
| Indicated Velocity Head | $h$ | 0.292397585 | 0.386375967 | 0.412719215 |
| Effective Wheel Radius | $R_E$ | 2.896453323 | 2.880986301 | 3.475743988 |
| Fill Profile Area | $A_{load}$ | 0.005970905 | 0.010187953 | 0.011683965 |
| Bucket Fill Volume | $V_f$ | 0.011374575 | 0.020184373 | 0.020477317 |
| Bucket Presentation Duration ( s ) | $t_f$ | 0.079136197 | 0.061081135 | 0.059957376 |
| Cycle Time ( s ) | $T$ | 7.597074927 | 6.596762626 | 7.674544109 |
| Bucket Fall Time ( Half-Cycle ) | $T_F$ | 3.798537464 | 3.298381313 | 3.837272054 |
| Angular Velocity ( cycles/second ) | $f$ | 0.131629609 | 0.151589508 | 0.130300899 |
| Angular Speed ( radians/s ) | $\omega$ | 0.827053223 | 0.952464969 | 0.818704697 |
| Fill Weight per Second | $K$ | 1410.000593 | 3241.65993 | 3350.346655 |
| Effective Fall | $H-h$ | 3.884519835 | 3.625343899 | 4.311265875 |

### Efficiency Metrics

| | | Hayes Mill | Coppice Mill | Stone Mill |
|---|---|---|---|---|
| Denny Spill Efficiency | $\varepsilon_D$ | 0.778934001 | 0.737196438 | 0.723424308 |
| Denny Combined Energy Efficiency | $\varepsilon_C$ | 0.080949233 | 0.056082689 | 0.049741959 |
| Retentional Efficiency Fraction | REF | 0.670564895 | 0.629184508 | 0.620193244 |
| Approximate REF | $REF_{approx}$ | 0.669976534 | 0.628663178 | 0.619643285 |

### Hydraulic Power Outputs

| | | Hayes Mill | Coppice Mill | Stone Mill |
|---|---|---|---|---|
| Torque due to Bucket Water Gravitation | $\tau_g$ | 3311.259248 | 6169.325086 | 8821.3951 |
| Torque due to Water Entry Impulse | $\tau_w$ | 0.020095731 | 9.256773667 | 0.576498802 |
| Power due to Bucket Water Gravitation | $P_g$ | 5477.17527 | 11752.13205 | 14444.2352 |
| Power due to Water Entry Impulse | $P_w$ | 0.016620239 | 8.81675264 | 0.471982277 |
| Denny Gravitational Power In | $P_{gD}$ | 8168.001805 | 18678.35571 | 23289.89448 |
| Denny Kinetic Power In | $P_{wD}$ | 412.2807679 | 1252.49949 | 1382.752443 |
| | | | | |
| Denny Simplified Power Input | $P_D$ | 8580.282573 | 19930.8552 | 24672.64693 |
| Denny-Warren Model Total Power | G | 5477.19189 | 11760.9488 | 14444.70718 |
| Borda Power | $P_B$ | 5683.315654 | 12378.38179 | 15135.61142 |
| Franklin Institute Power | $P_{FI}$ | 5044.896358 | 10927.6187 | 13386.98237 |
| | | | | |
| Gate Power ( watts ) | G | 5477.19189 | 11760.9488 | 14444.70718 |
| | | | | |
| Modelled Power Split ( $P_w/P_g$ ) | | 3.03445E-06 | 0.000750226 | 3.26762E-05 |

### Regional Constants

| | | Hayes Mill | Coppice Mill | Stone Mill |
|---|---|---|---|---|
| Water Density ( $kgm^{-3}$ ) | $\rho$ | 999.6445 | 999.6445 | 999.6445 |
| Acceleration Due to Gravity ( $ms^{-2}$ ) | $g$ | 9.8132693 | 9.8132693 | 9.8132693 |
| Specific Weight of Water | $\gamma$ | 9809.780683 | 9809.780683 | 9809.780683 |

### Geometrical Corollarys

| | | Hayes Mill | Coppice Mill | Stone Mill |
|---|---|---|---|---|
| Spill Angle | $\theta_{spill}$ | 0.425531464 | 0.278731268 | 0.259113294 |
| Angle of Exhaustion | $\psi_x$ | 2.162611013 | 2.065070562 | 2.034035917 |